Lecture Notes in Chemistry

Edited by G. Berthier, M. J. S. Dewar, H. Fischer,
K. Fukui, H. Hartmann, H. H. Jaffé, J. Jortner,
W. Kutzelnigg, K. Ruedenberg, E. Scrocco, W. Zeil

3

Svetozar R. Niketić
Kjeld Rasmussen

The Consistent Force Field:
A Documentation

Springer-Verlag
Berlin · Heidelberg · New York 1977

Authors

Svetozar R. Niketić
Department of Chemistry
Faculty of Science
University of Beograd
P.O. Box 550
YU-11001 Beograd

Kjeld Rasmussen
Chemistry Department A
Building 207
The Technical University of Denmark
DK-2800 Lyngby

Library of Congress Cataloging in Publication Data

Niketic, Svetozar R   1944-
    The consistent force field.

    (Lecture notes in chemistry ; 3)
    Bibliography: p.
    Includes index.
    1. Chemical equilibrium. 2. Matter--Properties.
3. Field theory (Physics) I. Rasmussen, Kjeld,
1936-   joint author. II. Title.
QD503.N54   541'.042   77-24235

ISBN-13: 978-3-540-08344-3   e-ISBN-13: 978-3-642-93063-8
DOI: 10.1007/978-3-642-93063-8

2152/3140-543210

PREFACE

The preface of a book is probably the pleasantest part to write, as it gives the authors an opportunity to express their gratitudes towards all those who have helped.

This book has grown out of two dissertations and two series of seminar notes. We are grateful to those students and colleagues who contributed valuable criticism. Two former students appear as co-authors of individual chapters.

One author (KjR) has spent, over the years 1969-74, several months at Chemical Physics Department, The Weizmann Institute of Science, Israel, studying and developing the CFF. Without the generous hospitality of Professor Shneior Lifson, the Department and the Institute the whole project would not have been.

The other author (SRN) has spent a total of three years at Chemistry Department A, The Technical University of Denmark, as a graduate student of KjR and Professor Flemming Woldbye. He was supported for the greater part of his stay by the Danish Natural Science Research Council. During that period the CFF version described in the book took shape. When Danish computer centres had to charge users for their services, all costs were met through grants from the same Council.

Travel grants were donated by two private funds, Tribute to the Danes through Scholarships in Israel (to Klavs Kildeby) and Berg's Fund for the Advancement of Danish Engineering Science (to KjR).

Mrs. Birgit Rasmussen composed the manuscript, using a text editing programme written at the Danish Data Archives and maintained by the Technical University Computing Centre.

A grant towards part of the cost of the machine editing of the manuscript was provided by G. A. Hagemann's Memorial Fund.

We wish to acknowledge the good service of the academic and technical staff of the Computing Centre throughout the years.

Dr. Ivan Gutman is thanked for valuable comments on the terminology of graph theory.

All drawings were made by Mrs. Rita Bloch Hansen.

Professor Flemming Woldbye is to be thanked for having fostered conformational calculations at Chemistry Department A, and for having established the contact between Professor Shneior Lifson and KjR, as well as between the two authors.

The Board of Chemistry Department A, through the Director, Professor N. Hofman-Bang, are thanked for having given us sufficient everyday facilities to carry out our project.

Last, but not least, we want to thank our wives for constant encouragement and prodding throughout the years.

Beograd and Copenhagen

in April 1977

Svetozar R. Niketić   Kjeld Rasmussen

CONTENTS

1 INTRODUCTION

Kjeld Rasmussen

This book deals with the Consistent Force Field, or rather with one specific realisation of ideas and methods developed by Professor Shneior Lifson of the Weizmann Institute of Science and his associates.

We undertook to write the book because a full documentation is needed by colleagues and students who want to apply and further develop the system. The documentation available until now is found in four Ph.D. Theses, one of which is in Hebrew, and in a large number of journal articles, some of which suffer from the requirement of laconic presentation. It is to be hoped that this book will serve as an exposition of the basis of the CFF as well as a presentation of the system developed at The Technical University of Denmark.

Noone can be reasonably well acquainted with a large programme without actually using it. Our system is available for distribution and can be installed by any experienced programmer. Experience has shown, however, that it is most fruitful to stay for a short while on the spot and learn how to use it.

Because the book will concentrate on this specific implementation, we think it should be preceded by a summary of what the CFF is and an overview of how it was developed. Rather than giving an exhaustive review, we have selected a limited number of key papers, which we present with some comments.

1.1 What the CFF is

The Consistent Force Field is a concept with which we try to bridge the gap between several theoretical and experimental techniques. There is nothing revolutionary about it, and it can be stated in very conventional terms:

(1) choose a model of the structure of matter;

(2) find a mathematical expression for the model;

(3) derive the numerical value of any quantity you might want to know;

(4) compare the calculated with the measured value and improve the model to obtain a better fit.

The special characteristics of the CFF lie in points (3) and (4), as will become clear later.

(1) For the model, we use the Born-Oppenheimer separation and drop all electronic motion. We then assume that all interatomic interactions are additive and mutually independent.

(2) This means that we may formulate the potential energy of any system as a sum. We further split the interactions into types, following any of the many schemes. One very simple example is

$$V = \sum_{\text{bonds}} 1/2\ K_b (b-b_o)^2 + \sum_{\text{angles}} 1/2\ K_\theta (\theta-\theta_o)^2 + \sum_{\text{torsions}} 1/2\ K_\phi (1+\cos n\phi) +$$

$$\sum_{i>j} \left(A_{ij} / d_{ij}^{12} - B_{ij} / d_{ij}^{6} + e_i e_j / d_{ij} \right)$$

but we can use any type of potential energy functions.

The initial values of the parameters K_b, b_o, K_θ, ..., B, e are taken from related work or are guessed.

It is important to emphasise that K_b etc. are not force constants. They are parameters of energy functions, just as b_o etc., A, B, and e. The force constants we use are derived numerically.

(3) We can now calculate the potential energy of any atomic arrangement, provided we have a set of values of the energy parameters, a description of the topology of the arrangement, and a set of atomic coordinates. The energy will have minima at those points in configurational space which correspond to equilibrium conformations for the chosen set of energy functions. The curvatures in these points are the force constants of vibrational motion around the equilibria. Let us expand the energy in a Taylor series around one of the minima:

$$V(\underline{r}) = V(\underline{r}_o) + \sum_i (\partial V/\partial r_i)_o \delta r_i + 1/2 \sum_{i,j} (\partial^2 V/\partial r_i \partial r_j)_o \delta r_i \delta r_j + R$$

The first term is the equilibrium energy. Its absolute value may or may not be physically significant, according to the choice of energy functions, but the relative energies of different equilibrium conformations of the same molecule are very significant in determining which conformation is preferentially taken.

The second term vanishes at equilibrium, as a necessary condition for equilibrium is that the gradient vanishes.

The third term represents the energy of vibration, and here we see that in our method the force constants are derived for each individual interaction between two coordinates.

The quantities we want to know may not first of all be the energies, but the equilibrium conformations. They are found as those sets of coordinates that minimise the energy. We shall not go into detail about this, but just mention that we have three quite powerful and general minimisation methods.

The most sophisticated of these uses the second partial derivatives of the energy, and we calculate them through the analytical formulae. This means that we have access to all force constants at the equilibrium conformation. It is now an easy task to calculate normal frequencies and modes of vibration of a molecule in its equilibrium state.

Any other property that depends on these basic static and dynamic ones can be calculated. Let us just mention crystal structure, thermodynamic functions, and infrared intensity and circular dichroism.

(4) The geometries and frequencies obtained may not be frightfully good approximations to the measured values. It all depends on the energy functions and parameter values chosen. Though we do have examples of very good fits even at this stage, the rational thing to do is to change the energy parameters automatically to give a better fit. This is not easy. We change those parameters we wish to optimise by a small amount, one at a time, and calculate the resulting change in the observable quantities. Then we can calculate the derivatives of observables with respect to energy parameters. The derivatives are used in a linearised least-squares algorithm to determine the optimal changes in parameter values.

Now we see that the most important result of a CFF study may not be a set of conformations or frequencies. It may rather be a set of parameters for a specific set of potential energy functions.

Let us now resume what the Consistent Force Field is.

It is a concept: choose a set of energy functions
choose a set of parameters
compute any observable
optimise the parameters by fitting calculated
to measured observables.

It is a method, or a collection of methods:
calculate equilibrium geometry through energy
minimisation
calculate vibrational frequencies at equilibrium
calculate any other property from those two basic
ones
repeat this for a set of molecules
optimise the energy parameters simultaneously
on all observed property values for the whole
set of molecules.

It is a vision: do what we have just indicated for sets of
related substances; in this way you build up
a set energy functions common to several classes
of substances which will represent faithfully
all possible static and dynamic data;
then you are able, with some confidence, to
make predictions for systems too large or too
complicated to be studied experimentally or
theoretically.

The CFF is thus purely empirical, though with theoretical under-
tones, in the sense that quantum chemical arguments and calculations
assist us in the choice of the analytical forms of potential energy
functions and in some cases also of initial parameter values. In
this sense, our method is one more way of putting quantum chemistry to work.

1.2 Background

Although not stated in the early CFF papers, the work was originally undertaken in order to extract quantities necessary for the
calculation of protein conformation.

In the mid-fifties, many theoretically inclined biochemists applied statistical-mechanical methods to the helix-to-coil trans-
itions in polypeptides. One of the estimated and widely used results
is the Lifson-Roig model (Lifson and Roig 1961), in which a number
of characteristics of a polypeptide chain was derived from the conformational partition function in the space of all Ramachandran
angles. In the following years, the method was developed to treat
DNA and other single- and double-stranded polynucleotides (Lifson
and Zimm 1963, 1964).

As the partition function is derived from the conformational potential energy, there was clearly a need for a better understanding of
this quantity, and a number of groups set out to obtain it. It
turned out to be a bigger task than anticipated, but a lot of new
insight developed in the process.

1.3 Pre-CFF

Lifson started with the classical example of conformational analysis, the medium-sized cycloalkanes (Bixon and Lifson 1967). We
have got here a system of molecules which is simple in that only two
types of atoms are involved; yet it is complicated through the
occurrence of strain energy, as all rings except cyclohexane are
strained. In addition, a wealth of chemical and physical data are
available for comparison with computational results.

The potential energy was split into a sum of terms in the internal
coordinates of bond lengths, valence angles and torsional angles
(see Section 1.1), and the energy was minimised through unconstrained movement of the atoms (see Chapter 5). Considering what is
available today, the programme was rather simple; yet very reasonable results on conformations and excess enthalpies came out. The
same programme was used to help solving the crystal structure of a
derivative of cyclodecane (Dunitz et al. 1967), and the early
methods and results were summarised by Lifson (1968). A variant of
the minimisation method was used for refinement of the crystal
structures of two proteins (Levitt and Lifson 1969).

1.4 The ascent of CFF

In 1968 the full CFF method was presented (Lifson and Warshel 1968).
In this well-known paper it is demonstrated how calculation of
equilibrium conformations, eigenfrequencies of vibration and excess
enthalpies, followed by least-squares fitting of energy function
parameters to make calculated approach measured observables, can be
used to develop a force field which is consistent in the sense that
it reproduces equally well 94 individual observables of 10 n-alkanes
and cycloalkanes.

This pioneering work was soon extended by incorporation of cross terms (Warshel and Lifson 1969), crystal structures and lattice vibrations (Warshel and Lifson 1970), anharmonicity (Warshel 1971) and heteroatoms in amides and lactams (Warshel, Levitt and Lifson 1970) and Pyrrolidones (Shellman and Lifson 1973). Summaries were presented by Lifson (1972, 1973).

1.5 In the wake of CFF

Until then, the potential energy functions chosen had been of the modified Urey-Bradley type: quadratic or Morse functions for bond terms, quadratic for valence and Pitzer for torsional angle terms, Buckingham or Lennard-Jones for non-bonded terms, and linear plus quadratic for geminal interactions.

A valence force field with various cross terms was developed for olefins (Ermer and Lifson, 1973, 1974) and very strained bicyclic systems (Ermer 1974).

The hydrogen bond in amide crystals was studied (Hagler, Huler and Lifson 1974) using a new optimisation technique (Hagler and Lifson 1974). An extensive review of calculations, not only CFF, on proteins, is in preparation (Hagler and Lifson, to be published).

A major development occurred when pi electron systems were treated with a self consistent semiempirical method, while retaining the purely empirical CFF for the sigma bonding (Warshel and Karplus 1972, 1974, Warshel 1973). This programme was also extended to crystals (Huler and Warshel 1974). From an application point of view, the essential new feature is that it is possible to calculate conformations of electronically excited states and vibronic inter-actions; even rotatory strengths may be calculated (Schlessinger and

Warshel 1974). A review has recently been announced (Warshel 1977).

As stated at the beginning of this Introduction, we do not give here an extensive bibliography of the CFF. Some additional developments and applications are discussed in Chapters 5 and 8.

2 THE PROGRAMMING SYSTEM

Kjeld Rasmussen and Svetozar R. Niketić

2.1 Introduction

The present version of the programming system is based on programmes developed at Chemical Physics Department, The Weizmann Institute of Science, Rehovot, Israel, prior to 1970. The original idea of adapting the existing programmes to handle also transition metal complexes developed into a project of writing a completely new version of the conformational programmes with the following objectives.

1. The programmes should be based on the consistent force field concept of Lifson (1968, 1972).

2. Without loss of their original flexibility the programmes should be designed to treat any type of transition metal compound and in particular octahedral, square planar and tetrahedral metal chelate complexes.

3. They should be written entirely in FORTRAN IV for the IBM Operating System/370.

4. They should be built in segments so as to maximise the ease of user extensions and modifications and to optimise the main and auxiliary storage management and performance.

Many elegant computational details from the original version were retained, but numerous modifications were introduced, and some critical sections (mainly the energy minimisation routines) were entirely replaced by new versions. The normal coordinate analysis and the parameter optimisation are totally new constructions.

2.2 Outline of the programming system

For the sake of convenience in giving a review of the programmes that constitute the present version of our programming system for CFF calculations, we will classify them into seven sections (Table 2.1).

2.2.1 Section I

This section consists of the main programme (MAIN) and the sub-routine TID. The main programme controls the entire CFF computation according to the supplied global control parameters which define (1) specification of the force field type, (2) the number of molecules that are to be treated simultaneously, (3) where to find the energy function parameters, (4) whether to save the atomic coordinates of the equilibrium conformations on disk files, and (5) the number of cycles of energy parameter optimisation.

Subroutine TID is a small subroutine which prints messages about the elapsed and cumulative time when referenced from various parts of the programming system.

2.2.2 Section II

This section contains the programme NPAR which is used for processing of parameters for the potential energy functions.

Its modes of operation are the following: (1) reading of energy parameters from card images in the input stream and assignment of energy parameters and of internal control parameters to the corresponding arrays; (2) creation of a data set of energy parameters; (3) reading of parameters from a previously, created data set and sorting them as above; (4) updating of parameters in a permanent data set by values read from card images; (5) updating of selected parameters after each cycle of optimisation.

<u>Table 2.1</u>

Sections and subroutines

Section I	Section IV	Section V
MAIN	CONFOR	MOLEC
TID	REFXYZ	DECODE
	TESTER	LENGTH
Section II	INTOUT	COSTHE
NPAR	ORTOUT	DIFBON
	MONSTR	DIFANG
Section III	CHARGP	MATPAK
BRACK	DIPOLE	BONDP
METALS	STEEPD	EBOND
CODER	DAVID	BFUNC
FCODE	GAUSS	THETAP
FCODE1	STEPSZ	ETHETA
SELECT	CHLSKY	TFUNC
SELEQT	LINSOL	PHIP
REDUCE		EPHI
HATOMS	Section VI	PFUNC
SIDEAT	VIBRAT	MATFOR
MATR	EIGEN	UREYP
TRANS	INTENS	EUREY
MATR2	SYMANA	UBFUNC
MATR3		NBONDP

MKLIST	Section VII	ENBOND
ENCODE	OPTIM	NBFUNC
	RDEXP	
	BUILDY	
	BUILDZ	
	ZMATRX	
	LSTSQR	

2.2.3 Section III

This section contains the programmes for topological analysis of molecular structures and for building molecular geometries (calculation of cartesian atomic coordinates) on the basis of specially coded molecular formulae (line formulae). The programmes yield atomic coordinates and lists of interactions and non-bonded exclusions for each molecule. Optionally, one can enter the cartesians (obtained from previous calculations or from other sources) or specify particular conformations by entering values of torsional angles. The programmes of this section are the following:

BRACK is the control programme. It also reads and analyses the line formulae.

METAL assigns atomic symbols and atomic weights for metal atoms in coordination compounds.

CODER, with functions FCODE, FCODE1, SELECT, and SELEQT, perform the topological analysis of molecular structures.

REDUCE calculates molecular geometry and outputs cartesian atomic coordinates. Molecules are constructed on the basis of an internal library of standard bond lengths and valence angles using the topological information carried over from BRACK and CODER.

HATOMS completes a structure by adding hydrogens on chain atoms of a structure obtained from an X-ray diffraction study.

SIDEAT adds sideatoms on chain atoms and is used by REDUCE and HATOMS.

A number of common operations from matrix algebra used in constructing and transforming the coordinates is contained in subroutines MATR, MATR2, MATR3 and TRANS.

MKLIST prepares lists of interactions and non-bonded exclusions which are packed into integer words, using the function ENCODE, and written on to a disk file to be used later by programmes of the following sections.

2.2.4 Section IV

This section contains the programmes for conformational analysis. All calculations on conformations are governed by CONFOR. According to the input information supplied to and saved by BRACK for each molecule, CONFOR will select the minimisation method(s) to be employed; it will print the total energy and its distribution on bonds, angles, torsions etc. of the initial and the final conformations, as well as the energy minimisation history and the final energy gradient; and it will save the final coordinates on a disk file.

The following programmes are controlled by CONFOR.

TESTER calculates numerical first and second-order partial derivatives of energy with respect to cartesians and compares them to the corresponding analytically computed values, which are normally used in all calculations. It can print complete tables of derivatives or only messages about those derivatives found to be in error, together with the numerical and analytical values. TESTER is used only occasionally when developing and testing new formulae for potential energy calculations.

REFXYZ transforms the coordinates after minimisation in order to ease the comparison of the structures before and after minimisation. Coordinates are transformed to a molecular system defined by three atoms: one in the origin, one on the X axis, and one in the XY plane.

INTOUT prints lists of internal coordinates before and after minimisation, and a list of cartesian atomic coordinates for each molecule at the end of the computation.

ORTOUT prepares card image output of cartesian coordinates and other information, to be used by the standard plotter programme ORTEP II.

MONSTR works analogously, providing input for another plotter programme MONSTER.

CHARGP assigns partial charges to the chain atoms according to the list from NPAR, and to the sideatoms through electroneutralisation of groups. Molecular gross charge (of positively or negatively charged complex ions) is currently distributed evenly on amine hydrogens or, in the absence of these, on non-hydrogen sideatoms.

DIPOLE calculates molecular dipole moments from atomic coordinates and fractional charges.

STEEPD performs energy minimisation by the method of steepest descent.

DAVID performs energy minimisation with the Davidon-Fletcher-Powell algorithm.

GAUSS is a programme for energy minimisation that uses a modified Newton method based on the Cholesky decomposition of the matrix of second derivatives.

STEPSZ finds the optimal stepsize for a given search direction found by DAVID or GAUSS.

CHLSKY performs the Cholesky factorisation of a symmetric positive definite matrix.

LINSOL solves a set of linear equations.

2.2.5 Section V

This section contains the programmes for calculation of the intramolecular potential energy due to different types of interaction and, if required, the first and second derivatives of the total energy with respect to cartesians. There are five analogous sets of subroutines which process the bond, angle, torsional, Urey-Bradley (as an option), and non-bonded terms.

In addition, this section contains the subroutines MATFOR and MATPAK used for processing of the derivatives of torsional angles and for packing of the matrix of second derivatives. Also, programmes DIFBON and DIFANG, which calculate derivatives of internals (bonds and angles) with respect to cartesians, are found here.

Finally, the section contains small function subprogrammes LENGTH and COSTHE, which compute distances and angles, and the subroutine DECODE for unpacking of integer words (see ENCODE in Section III).

All functions of the whole section are controlled by the programme
MOLEC.

2.2.6 Section VI

This section contains the programmes for vibrational analysis.
Subroutine VIBRAT performs a mass-weighting of the matrix of second
derivatives, and solves the eigenvalue problem of vibration in
cartesian space. This is done by subroutine EIGEN, which uses
Householder tridiagonalisation and a QR algorithm; it can provide
both eigenvalues and eigenvectors. VIBRAT derives the normal
frequencies from the eigenvalues, and may transform the eigenvectors
into de-massweighted cartesian or internal displacement coordinates.

Subroutine SYMANA analyses the normal coordinates expressed in terms
of internal displacement coordinates according to contributions from
changes in all bonds, angles and torsions. This greatly facilitates
symmetry assignments of normal modes.

Subroutine INTENS gives a crude classical estimate of infrared
intensities, from cartesian displacement coordinates and fractional
charges.

2.2.7 Section VII

Subroutine OPTIM controls the process of optimisation. RDEXP reads
from cards values of experimental data and uncertainties, counts
them and stores them on a background file. BUILDY calculates
weighted differences between calculated and experimental values.
BUILDZ calculates the elements of the Z-matrix of derivatives of
observables with respect to energy parameters, and ZMATRX puts the
matrix together. LSTSQR performs the least-squares algorithm.

2.3 Other programmes

2.3.1 Utilities

In addition to the system of programmes for conformational calculations we have constructed a number of small FORTRAN and JCL programmes for auxiliary data manipulation and maintenance of the system. Some FORTRAN programmes are used, for example, to copy, list and edit atomic coordinates stored unformatted on disk files.

2.3.2 CFFPLOT

This programme was written mainly to facilitate the development of potential energy functions. It uses the standard conventions of the CFF system for energy functions and their parameters, and can plot all types of interaction functions.

The plot formatting programme called by CFFPLOT was written by Mr. Niels Sondergaard.

2.3.3 CRYSTAL

The original version of the programme CRYSTAL, which was kindly provided by Professor John A. Schellman (University of Oregon, Eugene, Oregon), had the following functions: (1) calculation of cartesian atomic coordinates from unit cell data and fractional coordinates of a crystal structure; (2) generation of symmetry connected parts of molecules on two levels; (3) generation of helices; (4) translation and rotation of cartesian coordinates; (5) calculation of any bond length, or non-bonded distance, valence angle and torsional angle.

We have modified the programme slightly and added the possibility of calculating uncertainties in distances and angles and in cartesian coordinates from standard deviations in fractional coordinates as reported by crystallographers. Uncertainties are necessary for the weighting process used in optimisation.

2.3.4 EDITOR

This programme is written by Dr. Basil Meyer, Department of Physics, The Weizmann Institute of Science, and is maintained by its author in collaboration with Dr. Stephen Druck, Weizmann Institute Computer Center. It is a very versatile file handling system, which has proven indispensable in our programme development. The programme is written in assembler code for the IBM 370 system.

2.3.5 ORTEP

This is the standard crystallographic programme ORTEP II from Oak Ridge National Laboratory, written by Johnson (1965).

2.3.6 MONSTER

This is a cheap, yet versatile FORTRAN plotter programme with a particularly easy graphic input language. It is written and maintained by Dr. Per Jacobi, Laboratory of Datalogy, School of Architecture, The Royal Danish Academy of Fine Arts.

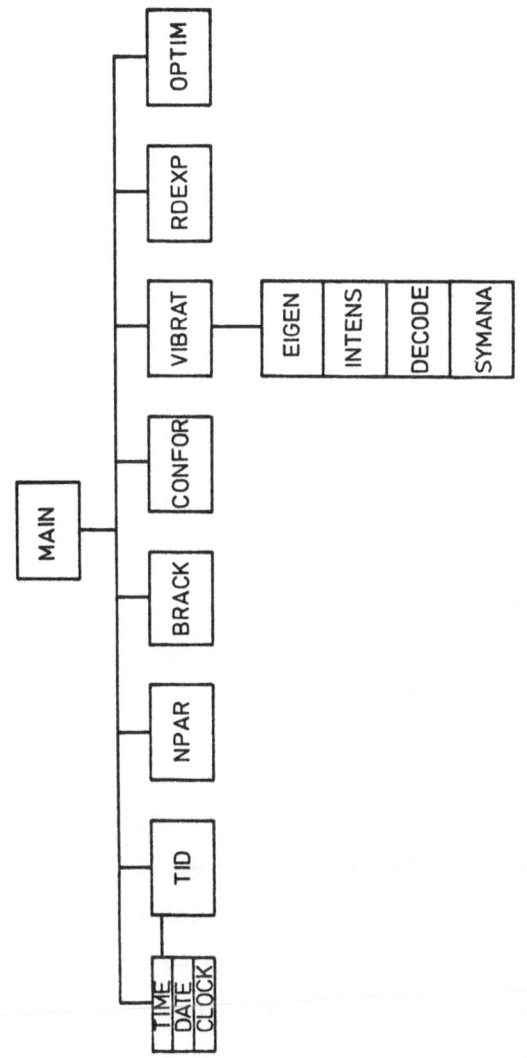

Figure 2.1. Main Overlay

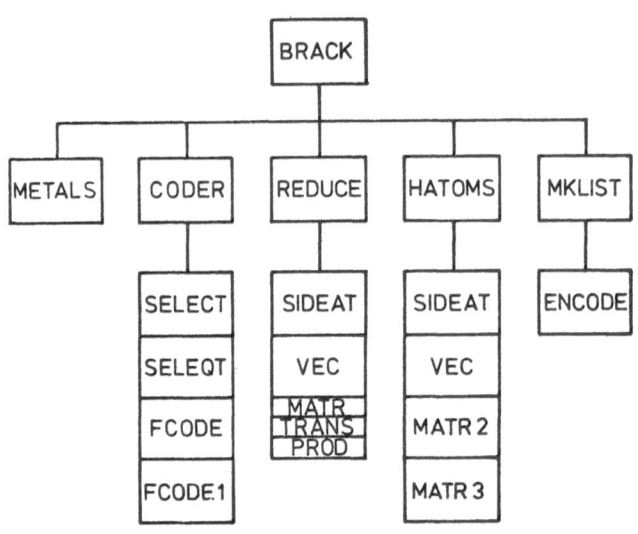

Figure 2.2. Overlay of Section III

2.4.2 JCL procedures

As the system is intended to be used by different people for many different purposes, it has to be made easily accessible. Subject to this requirement, it has to be as economical as possible. We have therefore written a series of JCL (Job Control Language) procedures, operating on a set of libraries. Most ordinary jobs can be run with just an EXEC card, specifying the permanent or temporary data sets for parameters, coordinates etc. chosen by the individual user.

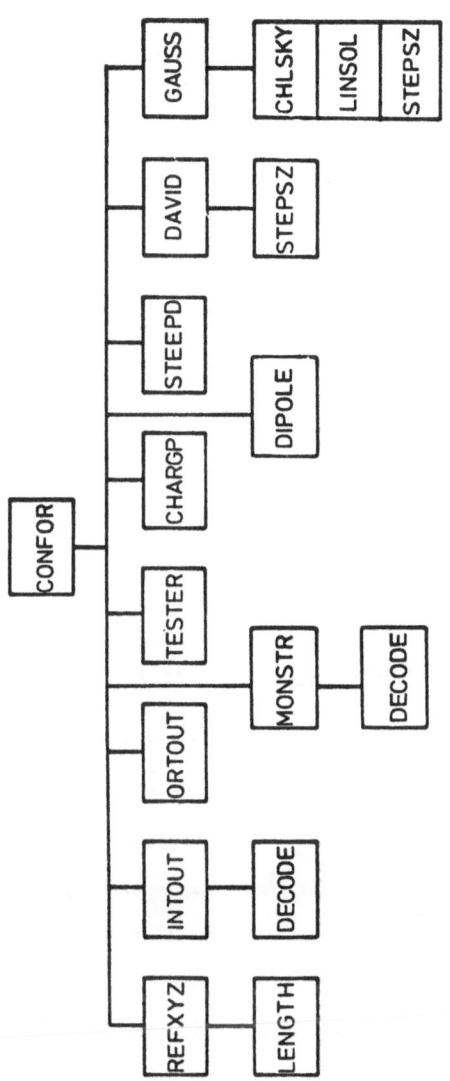

Figure 2.3. Overlay of Section IV

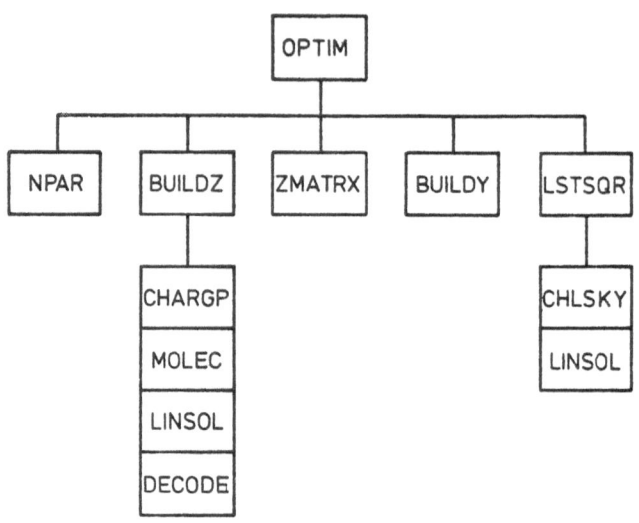

Figure 2.4. Overlay of Section VII

All the other necessary JCL is contained in the procedures. Compiler and linkage editor are called with parameters set to give minimum printout compatible with good sense. Temporary data sets are optimised with respect to blocksize and space pertaining to a small number of medium-sized molecules in a job, and both extensions and release of superflous space is done automatically. Channel separation is used where appropriate. The programmes may be equally well run from cards and by remote operation using data sets of card images.

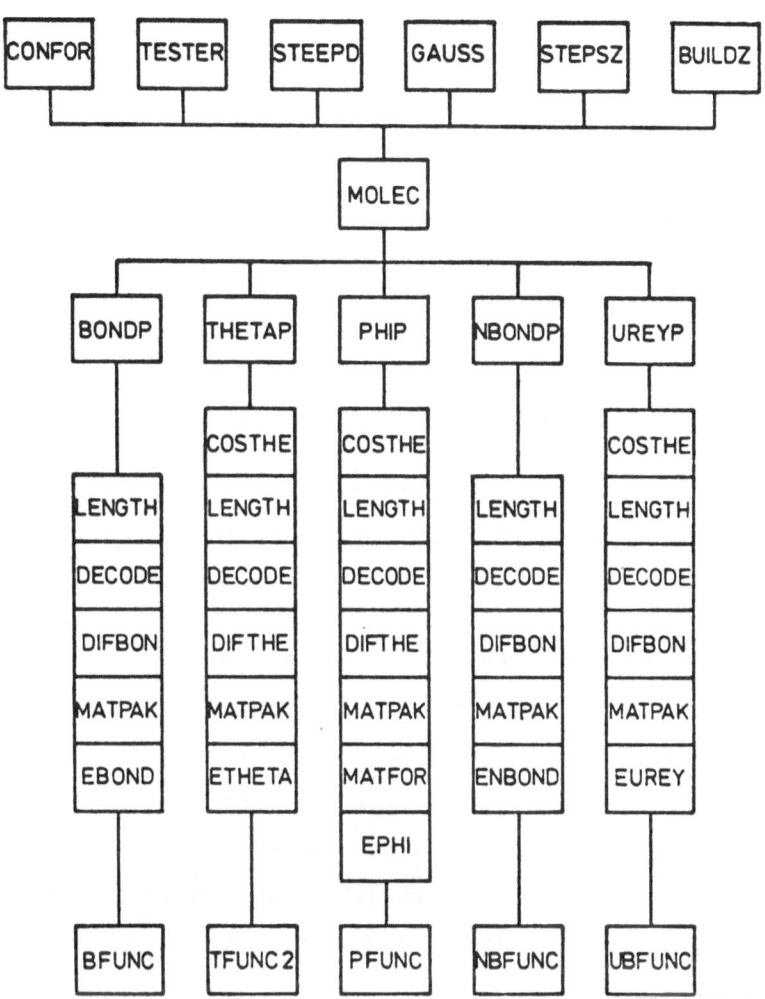

Figure 2.5. Overlay of Section V

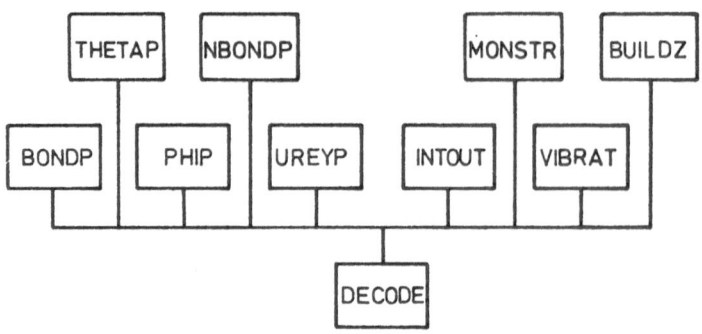

Figure 2.6.   Calls of DECODE

In what follows, we shall comment on the structure and use of four
catalogued procedures, shown diagrammatically in Figures 2.7 - 2.10.

Some symbols are common to all figures. SCREEN is any remote
terminal. In our Department we use an Infoton Vistar operating on a
1200 baud line. WITS (Waterloo Interactive Terminal System) is the
principal file handling system of NEUCC. It is installed as a subset
of TSO (Time Sharing Option), access to which is not public, and
corresponds roughly to the edit mode of TSO. HASP (Houston Automatic
Spooling Program) is the job scheduling and accounting system, and
OS is the IBM Operating System 370. JOB FILES are the user's private
data sets containing card images for job set-up and input.

2.4.2.1 CFFCLIB

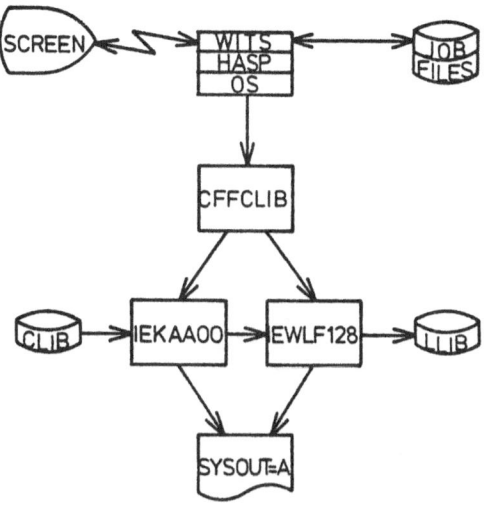

Figure 2.7. CFFCLIB

This procedure reads card images from a source text in CLIB into a work file and translates it with the FORTRAN H compiler IEKAAOO. The object code is transferred with the linkage editor IEWLF128 to LLIB. Line printer output is limited to compiler statistics and a map of LLIB.

2.4.2.2 CFFG

This is intended for routine runs with the precompiled programme in LLIB. Input cards to the linkage editor must be present as a member of CLIB. Permanent files for parameters and coordinates are specified on the // EXEC card. The remaining about twenty temporary files need no external specification. Units 5 and 25 must refer to card images, while units 10 and 20 are used unformatted. Unit 5 is the ordinary input file, and unit 25 is used only for input of experimental data for optimisation, which may be bulky; in this way they may be stored in a separate data set, which may be a tape file.

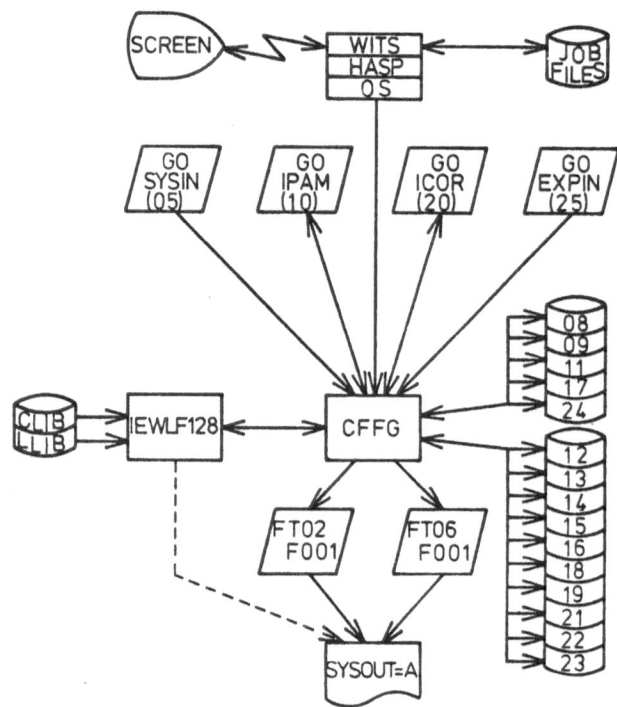

Figure 2.8. CFFG

Output appears on units 2 and 6, which are both routed to the lineprinter. Alternatively, unit 2 may be dummied, thus quenching the output from Section II of the programming system, which will mostly be identical in routine runs on the same set of molecules.

2.4.2.3 CFFCLG

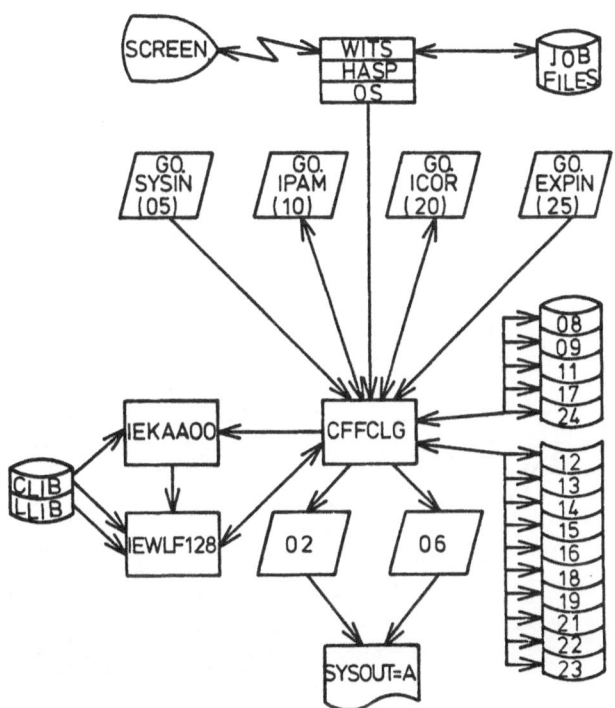

Figure 2.9. CFFCLG

This is intended for programme development, as is the following one, CFFECLG, and it contains one feature in addition to those of CFFG. CLIB contains a member, consisting of one to a few subprogrammes, which are translated with FORTRAN H and overwritten on the corresponding programmes in LLIB.

2.4.2.4 CFFECLG

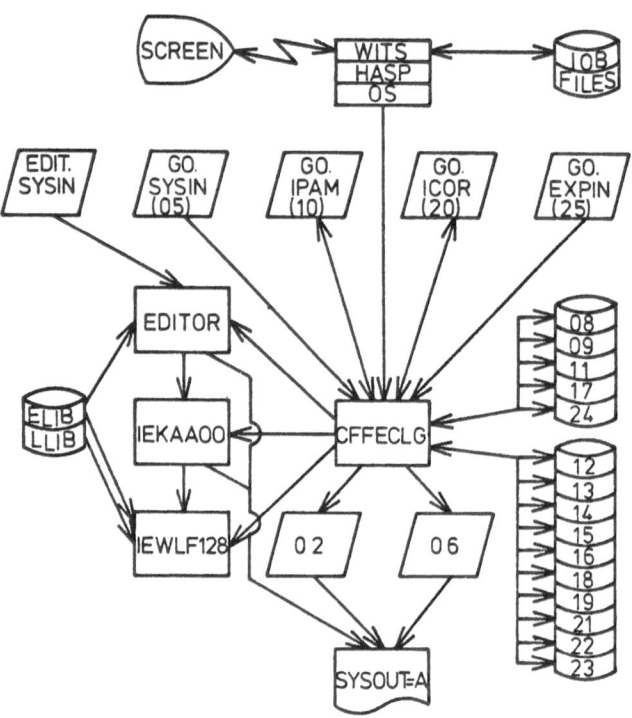

Figure 2.10. CFFECLG

With this procedure another way of modifying the programme is used. A member of ELIB, which takes the place of CLIB in the other procedures, is handled with the programme EDITOR (see Section 2.3.3).

The edited programme is translated and printed, whereupon the procedure performs the same functions as the others.

2.4.2.5 Listing of CPPECLG

As guide for the prospective user, we list the JCL procedure as it is used at the time of going to press. It will give the experienced programmer a notion of the resources required by the programme system.

```
//*A108001 USER=KJELD TLF=3334&3368
//*A108001P,NEU ,'R A S M U S S E N  ',TUESDAY   24.08.76 , 13.20.
//CPPECLG PROC CLIB='NEU.A108001.SOURCLIB',
//             LLIB='NEU.A108001.LINKLIB',
//             ELIB='NEU.A108001.SOURCLIB',EMEM=CORINA,
//             PARLIB='NEU.A108001.PARAM',
//             CORLIB='NEU.A108001.KOORD',
//             INPUTS='NEU.A108001.INPUTLIB',
//             INPUT=DUMMY,EXPIN=DUMMY,
//             NEWSLIB='NEU.A108001.KJELDJOB',NEWS=NEWS,
//             PAM=RESERVE,COR=RESERVE,ORT=DUM,MST=DUM,
//             COPTION='NOSOURCE,NOMAP',
//             LMEM=NADIA1,LOPTION='NOLIST,NOMAP,OVLY',GR=400K,
//             IPRINT='UNIT=AFF=ASYSPRT',NPRINT='UNIT=AFF=ASYSPRT',
//             EPRINT='UNIT=AFF=ASYSPRT',CPRINT='UNIT=AFF=ASYSPRT',
//             LPRINT='UNIT=AFF=ASYSPRT',
//             LOG2='UNIT=AFF=ASYSPRT',LOG6='UNIT=AFF=ASYSPRT'
//*           WRITTEN ON 29 MAR 76 BY KJELD
//*      NEWS
//NEWS    EXEC PGM=EDITOR,PARM=SS,REGION=64K
//STEPLIB   DD DSN=NEU.A108001.RUN,DISP=SHR
```

```
//&SYSPRT   DD UNIT=PRINT

//SYSPRINT  DD DUMMY

//SYSUT1    DD DSN=&NEWSLIB(&NEWS),DISP=SHR

//SYSUT2    DD &NPRINT,DCB=RECFM=FBA

//SCRATCH   DD UNIT=SYSDA,SPACE=(TRK,(1,1))

//SYSIN     DD DSN=&NEWSLIB(NEWSOUT),DISP=SHR

//*      INPUT

//INPUT  EXEC PGM=EDITOR,PARM=SS,REGION=64K

//STEPLIB   DD DSN=NEU.A108001.RUN,DISP=SHR

//&SYSPRT   DD UNIT=PRINT

//SYSPRINT  DD DUMMY

//SYSUT1    DD DSN=&INPUTS(&INPUT),DISP=SHR

//SYSUT2    DD &IPRINT

//SCRATCH   DD UNIT=SYSDA,SPACE=(TRK,(1,1))

//SYSIN     DD DSN=&NEWSLIB(INPUTOUT),DISP=SHR

//*      EDIT

//EDIT   EXEC PGM=EDITOR,PARM=SS,REGION=64K

//STEPLIB   DD DSN=NEU.A108001.RUN,DISP=SHR

//&SYSPRT   DD UNIT=PRINT

//SYSPRINT  DD &EPRINT

//SYSUT1    DD DSN=&ELIB(&EMEM),DISP=SHR,LABEL=(,,,IN)

//SYSUT2    DD UNIT=SYSDA,SPACE=(TRK,(3,1)),DISP=(NEW,PASS),

//            DSN=&TRANSFER

//SCRATCH   DD UNIT=SYSDA,SPACE=(CYL,(3,1))

//SYSIN     DD DDNAME=IN

//*      FORT

//FORT   EXEC PGM=IEKAA00,REGION=256K,PARM='&COPTION',

//            COND=(4,LT,EDIT)

//&SYSPRT   DD UNIT=PRINT

//SYSPRINT  DD &CPRINT
```

```
//SYSPUNCH   DD DUMMY

//SYSUT1     DD UNIT=SYSDA,SPACE=(TRK,(3,1))

//SYSLIN     DD DSN=&LOADSET,DISP=(NEW,PASS),UNIT=SYSDA,
//              SPACE=(TRK,(3,1)),DCB=BLKSIZE=3200

//SYSIN      DD DSN=&TRANSFER,DISP=(OLD,DELETE)

//*       LKED

//LKED     EXEC PGM=IEWLF128,REGION=192K,
//              COND=((4,LT,EDIT),(4,LT,FORT)),
//              PARM='&LOPTION,SIZE=(184K,60K)'

//ASYSPRT    DD UNIT=PRINT

//SYSPRINT   DD &LPRINT,DCB=RECFM=FBA

//SYSLIB     DD DSNAME=SYS1.FORTLIB,DISP=SHR
//           DD DSN=&LLIB,DISP=SHR
//           DD DSNAME=SYS2.FORTLIB,DISP=SHR

//SYSUT1     DD UNIT=SYSDA,SPACE=(TRK,(57,19)),VOL=SER=MVTWK1

//SYSLMOD    DD DSN=&CFFSET(CFF),DISP=(NEW,PASS),UNIT=SYSDA,
//              SPACE=(CYL,(3,1,1),RLSE),VOL=SER=MVTWK2

//SYSLIN     DD DSN=&LOADSET,DISP=(OLD,DELETE)
//           DD DSN=&CLIB(&LMEM),DISP=SHR

//*       GO

//GO       EXEC PGM=CFF,REGION=&GR,
//              COND=((4,LT,EDIT),(4,LT,FORT),(4,LT,LKED))

//STEPLIB    DD DSN=&CFFSET,DISP=(OLD,PASS),VOL=SER=MVTWK2,
//              UNIT=SYSDA

//ASYSPRT    DD UNIT=PRINT

//FT05F001   DD DDNAME=SYSIN

//FT06F001   DD DDNAME=LCG6

//FT02F001   DD DDNAME=LOG2

//FT08F001   DD UNIT=SYSDA,DCB=(RECFM=VBS,LRECL=1612,BLKSIZE=1616),
```

```
//              SPACE=(TRK,(1,1))
//FT09F001  DD UNIT=SYSDA,DCB=(RECFM=VBS,LRECL=1612,BLKSIZE=1616),
//              SPACE=(TRK,(1,1))
//FT10F001  DD DSN=&PARLIB(&PAM),DISP=SHR,LABEL=(,,,IN)
//FT11F001  DD UNIT=SYSDA,DCB=(RECFM=VBS,LRECL=492,BLKSIZE=496),
//              SPACE=(TRK,(1,1))
//FT12F001  DD DCB=(RECFM=VBS,LRECL=3152,BLKSIZE=3156),
//              SPACE=(TRK,(2,1)),UNIT=SYSDA,SEP=FT11F001
//FT13F001  DD UNIT=SYSDA,DCB=(RECFM=VBS,LRECL=X,BLKSIZE=13030),
//              SPACE=(TRK,(10,3),RLSE)
//FT14F001  DD UNIT=SYSDA,DCB=(RECFM=VBS,LRECL=X,BLKSIZE=13030),
//              SPACE=(TRK,(19,19),RLSE)
//FT15F001  DD DCB=(RECFM=VBS,LRECL=X,BLKSIZE=13030),UNIT=SYSDA,
//              SPACE=(TRK,(38,19),RLSE)
//FT16F001  DD UNIT=SYSDA,DCB=(RECFM=VBS,LRECL=4249,BLKSIZE=4253),
//              SPACE=(TRK,(2,1))
//FT17F001  DD DCB=(RECFM=VBS,LRECL=2055,BLKSIZE=2059),
//              SPACE=(TRK,(1,1)),UNIT=SYSDA,SEP=FT15F001
//FT18F001  DD UNIT=SYSDA,DCB=(RECFM=VBS,LRECL=1612,BLKSIZE=1616),
//              SPACE=(TRK,(10,3),RLSE)
//FT19F001  DD UNIT=SYSDA,DCB=(RECFM=VBS,LRECL=1612,BLKSIZE=1616),
//              SPACE=(TRK,(38,19),RLSE)
//FT20F001  DD DSN=&CORLIB(&COR),DISP=SHR,LABEL=(,,,IN)
//FT21F001  DD UNIT=SYSDA,DCB=(RECFM=VBS,LRECL=1612,BLKSIZE=1616),
//              SPACE=(TRK,(1,1))
//FT22F001  DD UNIT=SYSDA,DCB=(RECFM=VBS,LRECL=1612,BLKSIZE=1616),
//              SPACE=(TRK,(38,19),RLSE)
//FT23F001  DD UNIT=SYSDA,DCB=(RECFM=VBS,LRECL=540,BLKSIZE=544),
//              SPACE=(TRK,(1,1))
//FT24F001  DD UNIT=SYSDA,DCB=(RECFM=VBS,LRECL=1612,BLKSIZE=1616),
```

```
//                SPACE=(TRK,(38,19),RLSE),SEP=FT19F001
//FT25F001   DD DDNAME=EXPIN
//FT27F001   DD DSN=NEU.A108001.&ORT,DISP=SHR
//FT37F001   DD DSN=NEU.A108001.&MST,DISP=SHR
//LOG6       DD &LOG6,DCB=RECFM=FBA
//LOG2       DD &LOG2,DCB=RECFM=FBA
//SYSIN      DD DSN=&INPUTS(&INPUT),DISP=SHR,LABEL=(,,,IN)
//EXPIN      DD DSN=&INPUTS(&EXPIN),DISP=SHR,LABEL=(,,,IN)
```

2.4.3 Input-output

The temporary files are given symbolic names throughout the programmes. The symbols are integers of, hopefully, mnemotechnic value. In BLOCK DATA, identification of the symbols with the usual FORTRAN units is made. A summary is shown in Table 2.2.

All input-output operations are summarised in the charts of Figure 2.11.

2.5 Input manual

A rather detailed set of instructions on how actually to operate the system has been organised in an input manual, which is being updated in parallel with further programme development.

The manual pertains both to programme and system input. As an example, Table 2.3 shows time, line, region and IO - requirements for representative tasks.

Table 2.2

ORGANISATION OF BACKGROUND MEMORY

Reference external	number internal	Contents	Produced by subroutines	Used in subroutines
8	IND	Atomic coordinates	BRACK	CONFOR
9	IUD		CONFOR	BUILDZ
10	IPAM	Energy parameters	NPAR	NPAR
11	ICTR	Control parameters for individual mo-lecules	BRACK	CONFOR BUILDZ VIBRAT
12	IORD	Packed words	MKLIST	CONFOR BUILDZ VIBRAT
13	IVIB	DD matrix	CONFOR	VIBRAT
14	IBMA	DSX matrix	CONFOR	VIBRAT
15	IBMX	D vector, DSX matrix, DD matrix	CONFOR	BUILDZ

16	ICON	Calculated observables	CONFOR	BUILDZ BUILDY
17	IEXP	Measured observables	RDEXP	BUILDZ BUILDY
18	IZMA	Z matrix	BUILDZ	ZMATRX
19	IEIV	Eigenvectors	EIGEN	VIBRAT
20	ICOR	Final atomic coordinates	CONFOR	BRACK
21	IFRQ	Calculated frequencies	VIBRAT	BUILDY
22	IFMA	DD- matrix	CONFOR	BUILDZ
23	INTE	Fractional charges	CONFOR	VIBRAT
24	IEGV	Mass-deweighted eigenvectors	VIBRAT	BUILDZ

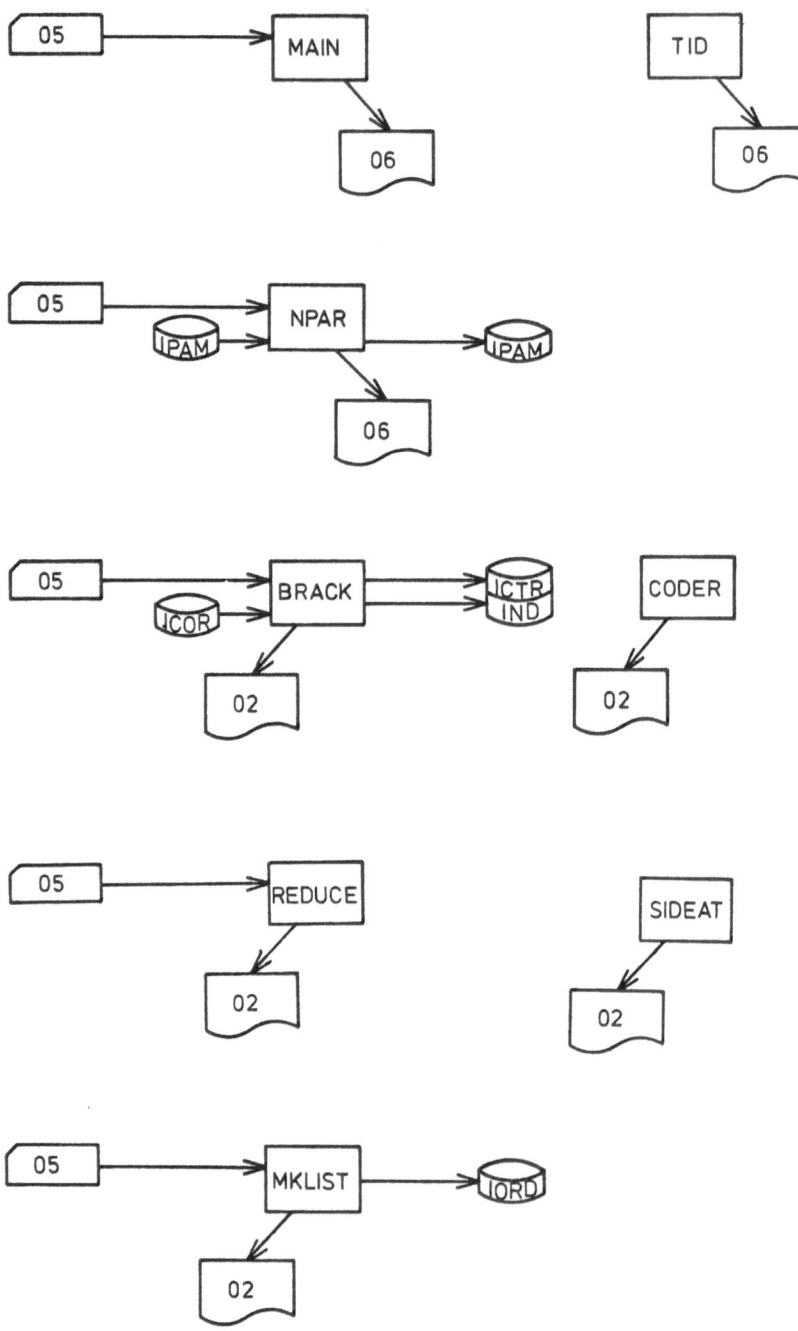

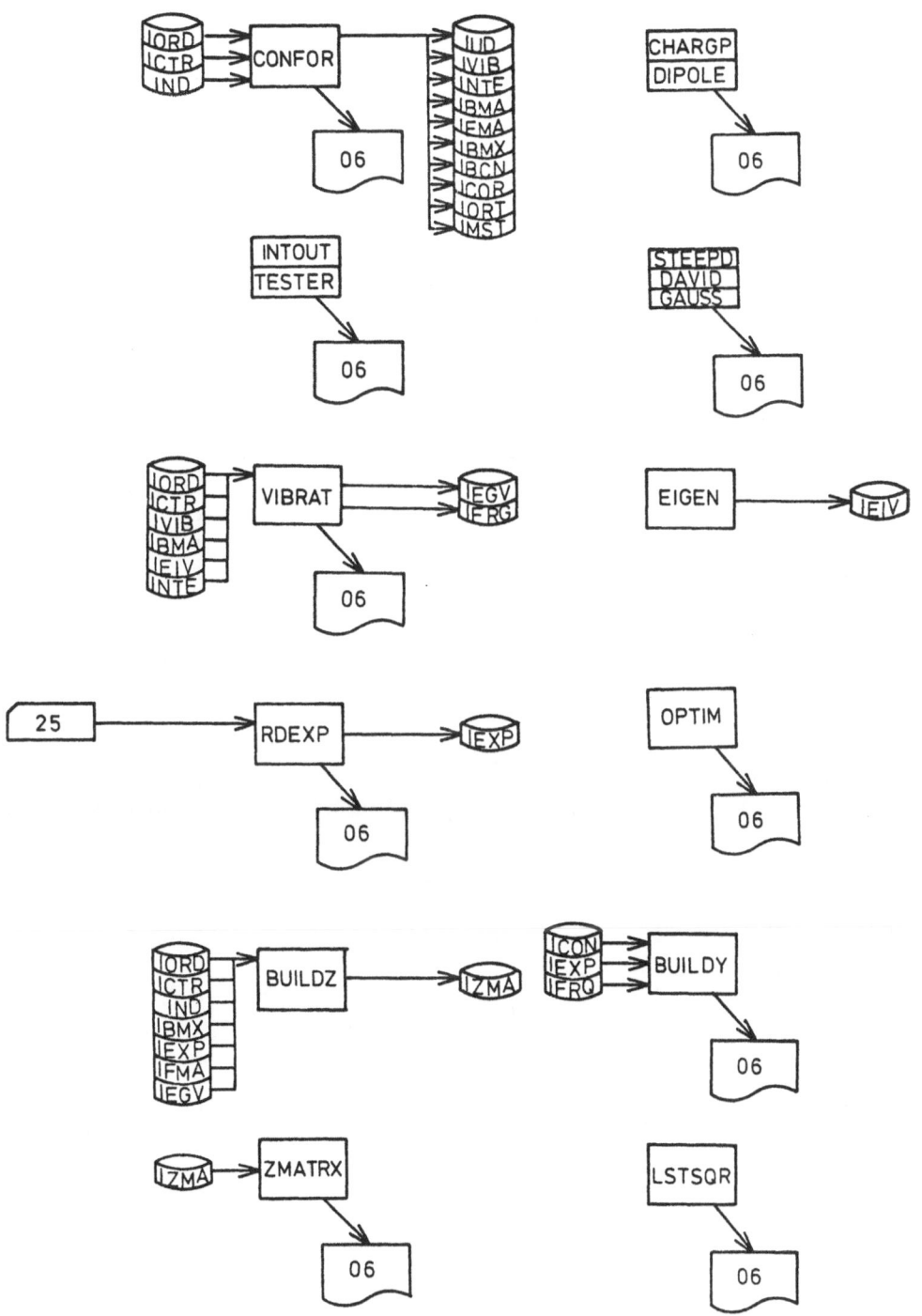

Figure 2.11. Input-output operations

Table 2.3. Examples of time, lines and core requirements

number of atoms	task	ttt seconds	111	REG K
12	full codings	2	800	340
24				340
46		3	800	340
64				340
12	initial conformation	1	200	360
24		1	250	360
46		1	300	360
64		2	450	360
12	minimisation: 10 steepest descent	1	200	360
24		4	300	360
46		10	350	360
64		20	500	360
12	minimisation: 10 Davidon	1	200	360
24			300	360
46		14	350	360
64			500	360
12	minimisation: 10 Newton	4	200	360
24		20	300	360
46		70	350	360
64		140	500	360

12	frequencies	1	100	380
24				380
46		10		380
64				380
12	normal coordinates (cartesians)	2	500	420
24				420
46		25	2000	420
64				420
12	normal coordinates (internals)	2	500	420
24				420
46		30	3500	420
64				420
12	numerical and analytical	8	1800	340
24	first and second derivatives			340
46				340
64				340
12	preparation for optimisation	6	200	400
24	on conformation	30	250	400
46			300	400
64			400	400

The examples were run with overlay.

3 MOLECULAR TOPOLOGY AND GEOMETRY

Svetozar R. Niketić and Kjeld Rasmussen

Three main topics will be considered in this chapter, each of them corresponding to one of three major computing steps preceeding the calculation and minimisation of molecular potential energy and, therefore, all other calculations under the programming system:

(a) analysis of molecular topology,

(b) generation of lists of interactions, and

(c) building of molecular geometry.

They are all performed by the programmes of section III of the system, controlled by programme BRACK.

Essentially, for a given molecular formula the programmes produce a set of cartesian atomic coordinates and lists of all intramolecular interactions. If the calculations include several molecules each of them is processed in turn, and the data (coordinates and lists of interactions) are stored sequentially on temporary disk files.

3.1 Molecular topology

In almost any case of computer application to the study of molecular structure and properties the first problem to be solved is that of communicating the initial structure information to the machine and of its internal representation suitable for further processing. This problem has been studied extensively from various aspects ranging from the very simple application in atom and bond numbering for the systematic listing of valence and torsional angles (Allen and Rogers 1969) to the large scale storage and retrieval of

structural information (Lynch 1968; Lynch et al. 1972) and to the
sophisticated systems for predicting routes in organic syntheses
(Corey 1971; Hendrickson 1971). In the sections that follow we shall
consider the problem of specifying the structural information to the
conformational programmes and the way the programmes build the
molecule from this information.

3.1.1 Topological representation of chemical structures

Apart from the use of graphical devices for direct input of
structural information (Corey and Wipke 1969) there are basically
two ways of representing chemical structures:

   (a) connectivity tables and matrices, and

   (b) linear notation.

Both may be called topological representations since they carry
essentially all information about the topological relations in a
molecular structure. The simplest form of a connectivity table is an
n-dimensional square binary matrix, the adjacency or atom-connec-
tivity matrix, which shows the precise arrangement in which the
atoms of an n-atomic molecule are connected. In addition, connectivity tables may contain the specification of unshared valence
electrons as in the be-matrices of Dugundji and Ugi (1973), or they
can be constructed as rectangular matrices containing details on
charges, masses and valencies of atoms and types, orders and other
properties of bonds at various levels of description (Lynch et al.
1972). A recent review was written by Gasteiger et al. (1974).

Widely used in chemical documentation are the linear notations which
consist of strings of characters and special symbols designed to
represent the molecular topology, bonds between atoms being implicit
in the sequence of symbols. An example is the familiar Wiswesser

line notation (Smith 1968).

We will now digress to the basic definitions of graph theory, which we have found particularly useful in rationalising and illustrating the transformation of molecular structure into linear notation used by the programmes.

3.1.2 From structural formula to linear notation

The application of graph theory to problems in structural chemistry is about a hundred years old (Cayley 1874), and so are the concepts of kenogram, plerogram and tree.

A linear graph is a set of vertices (or points or nodes) and a set of edges (or lines or branches) each of which joins two distinct vertices (Harary 1969; Marshall 1971). The analogy to molecular structure is obvious if we correlate vertices with atoms and edges with bonds. Acyclic molecules can, therefore, be represented by connected acyclic graphs called trees (Figure 3.1). Unambiguous mapping of any tree into a linear representation is always possible. Similarly, cyclic and polycyclic molecules can be represented by graphs having one or more cycles (Figure 3.2). All monocyclic and most polycyclic molecular structures yield planar graphs, which are graphs that can be drawn on surfaces, and in which no two of their lines intersect. According to a theorem of graph theory any of these graphs can be reduced to one of its spanning subgraphs by removal of one or more of its edges in such a way as to obtain a connected but acyclic graph, a spanning tree (Figure 3.3). The missing edges are called chords.

Figure 3.1 Ethane and its plerogram representation as a connected acyclic graph (tree). Graph elements a, b, c, d, e, f, g and h are nodes (vertices) corresponding to atoms, and elements 1, 2, 3, 4, 5, 6 and 7 are lines (edges) corresponding to bonds of a molecule.

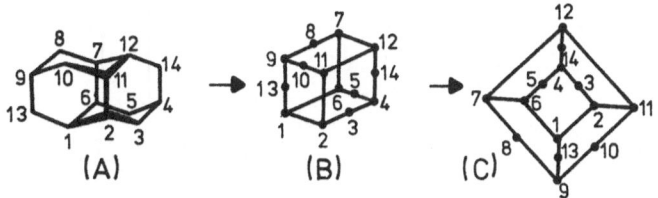

Figure 3.2 Kenogram representation of the pentacyclic molecule diamantane. Structure of molecular framework (A), non-planar graph (B), and planar graph (C) representing molecular topology.

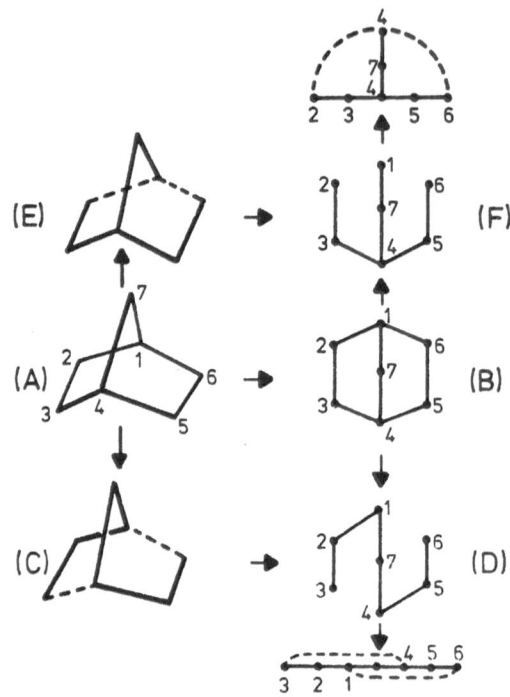

Figure 3.3 Kenogram representations of norbornane. Molecular (A), the corresponding cyclic graph (B), and two (E = F) and (C = D) of 17 possible ways of its reduction to a spanning tree.

Any molecular structure can thus be reduced to a linear notation through its spanning tree; and if the chords of the spanning tree are accurately labeled and incorporated into a linear notation, the latter becomes an unambiguous representation of a molecular topology.

Our system of notation provides all the necessary requirements for unambiguous topological representation of any single molecule, and molecular or ionic aggregate, such as a molecular complex or a unit cell.

When reducing the graph of a complex molecule it is mostly convenient to work with the kenogram (Figures 3.2 and 3.3), a representation in which sideatoms (see below) are left out, rather than with the plerogram (Figure 3.1), the representation of the full molecule.

3.1.3 Coding of formulae

We shall here give the rules for the coding of formulae that are to be processed by BRACK and CODER, and we shall show some illustrative examples. First we define some terms used in the discussion of the coding.

Chain atom is a vertex of an acyclic graph or spanning tree with degree > 1. Or it is any node of a kenogram. The degree of a vertex is the number of lines incident with it, for instance the valence of an atom in a molecule.

Sideatom is a node with degree equal to 1.

Chain is any path defined by a sequence of distinct chain atoms.

Sidechain is a chain starting on a chain atom in another chain in which case the latter chain is regarded as a chain of higher level. There may be several levels of sidechains.

3.1.3.1 Rules for coding line formulae

(1) A line formula consists of a string of selected alfameric characters and special symbols enclosed in parantheses.

(2) Single letters are used to represent atoms, resorting to the common chemical symbols wherever possible (Table 3.1). The present version of the programming system uses 12 symbols of which four and eight are predetermined as sideatoms and chain atoms, respectively, thus adding to each symbol its topological connotation. In this way, in a string of symbols all sideatoms between two adjacent chain atoms are assumed to belong to the former chain atom.

(3) To distinguish between different types of the atoms different letters are used, choosing among either visually similar letters (O or Q) or initial letters of different archaic names (N or A).

(4) Two consecutive chain atoms (with or without sideatoms in between) are assumed to be connected.

(5) If two or more sideatoms of the same type are carried by the same chain atom their total number is specified by a digit after the sideatom symbol, which is interpreted in the same way as a subscript in the normal chemical notation.

(6) Each sidechain is enclosed in parantheses and is written after the chain atom on which the branching occurs and also after any single sideatom attached to that chain atom.

(7) Absolute configuration is denoted by symbols R and S representing R and S of the nomenclature of Cahn, Ingold and Prelog (1966). The symbols are placed in front of the atomic symbols of chiral centres.

Tabel 3.1. Standard atomic symbols and codes

Atom	Symbol	Atom type number	Sideatom code
Sideatoms			
Hydrogen	H	1	1
Deuterium	D	2	1
Halogen	X	3	5
Oxygen, one bond	Q	4	4
Chain atoms			
Oxygen, two bonds	O	5	4
Nitrogen, trigonal (azote)	A	6	5
Carbon, trigonal (keto)	K	7	6
Carbon, tetrahedral	C	8	1
Nitrogen, tetrahedral	N	9	2
Metal,octahedral,tetrahedral or square planar	M	10	3
Sulfur (thio)	T	11	1
Phosphorus	P	12	1

(8) Any pair of identical symbols other than atomic symbols, absolute configuration designators and parantheses is used to indicate the pair of atoms effecting a ring closure, i.e. atoms corresponding to the vertices on a spanning tree of a molecular graph incident with a chord (removed edge). Although there is but two restrictions in the use of additional symbols, punctuation marks appear to be most convenient for checking purposes. The restrictions apply to the plus and minus signs, which are reserved for ionic charges in a coming version of the programme for ionic and molecular crystals.

3.1.3.2 Limitations

The present version of the conformational programme is designed to treat molecules with up to 67 atoms; versions for 12 and 99 atoms will be available.

This limit has been chosen because our main interest in the initial stages of programme application was concentrated on the conformations of tris-(diamine) and tris-(aminoacidato) metal chelate complexes of which the majority have less than 67 atoms. For example, the three complexes tris-(1,2-cyclohexanediamine) Cobalt(III) ion, tris-(2-amino-cyclohexane-1-carboxylato) cobalt(III) and tris-(phenyl-alaninato) cobalt(III) are of maximum size with respect to the dimensions of the present version of the programmes.

As a consequence, there are certain limitations in the current use of the linear notation, which are reflected in the total number of characters in the input formula (<300), the number of chain atoms (<30), and the number of chains (<20).

For the same reasons the other parts of the programming system are scaled to treat up to 80 bonds, 160 valence angles, 220 torsional angles and 1200 non-bonded exclusions per molecule.

Structures involving several hundred atoms (macromolecules) can be treated by other versions of the original programming system. An example is the coding of polypeptide chains using a superformula concept (Levitt and Lifson 1969; Levitt 1971).

3.1.3.3 Examples

Some examples of formulae coded for the conformational programme are shown below. Formulae for acyclic molecules illustrate the close resemblance to the common chemical notation:

n-butane   (CH3CH2CH2CH3)
1,2-ethanediamine   (NH2CH2CH2NH2)

Formulae for branched acyclic molecules show the use of the sidechain notation:

2,3-butanediamine   (CH3CH(NH2)CH(NH2)CH3)
or   (NH2CH(CH3)CH(CH3)NH2)
 -amino-isobutyric acid   (CH3C(CH3)(NH2)KQOH)
or   (NH2C(CH3)(CH3)KQOH)

Formulae exemplifying the chirality specification symbols:

(S)-alanine   (CH3SCH(NH2)KQOH)
or   (NH2SCH(CH3)KQOH)
meso-2,3-butanediamine   (NH2RCH(CH3) SCH(CH3) NH2)
(R)-lactic acid   (RCH(CH3)(OH)KQOH)
or   (CH3RCH(KQOH)(OH))

The use of connection symbols for simple cyclic structures is illustrated as follows:

cyclohexane   (,CH2CH2CH2CH2CH2,CH2)
(S)-proline   (.NHCH2CH2CH2.SCHKQOH)
or   (;CH2NHSCH(KQOH)CH2;CH2)

Extension to more complicated structures is straightforward:

adamanthane   (.CH2,CHCH2;CHCH2CH(,CH2)CH2.CH;CH2)
diamanthane   (,CH2ᴇCH;CHæCHCH2,CHCH2.CHᴇCHCH2C(;CH2)CH2æ.CH)

These examples show that it is possible to write a linear formula representing the same molecular structure in a number of different ways. In a ring structure, for example, we may take any two adjacent ring atoms, remove the bond between them, construct the linear formula from the resulting acyclic structure and attach the connection symbols to the atoms that were disconnected. For poly- cyclic molecules there are even more possibilities which may differ in the number of sidechains. There is no preference for any of the valid linear representations of a molecular structure although in some particular cases it may be advantageous to use the linear formula with the smallest possible number of sidechains.

3.1.4 Coordination compounds

In addition to the general rules outlined in the preceding para- graphs, the coding of formulae of coordination compounds requires the following special considerations.

Valid linear formulae representing a coordination compound are limited to those starting with a central metal atom as the first atomic symbol in the formula. Thus any line formula of a coor- dination compound should appear as follows:

(M......................)

All chelate rings are to be enclosed in parantheses. For example, a tris-(bidentate) structure will look as follows:

(M(.......) (.......) (.......))

Chirality symbols R and S placed before the symbol for a metal atom, M, signify the absolute configuration, related to the octahedron in such a way that R corresponds to Δ and S to Λ, where Δ and Λ are defined according to the Nomenclature of Inorganic Chemistry (1971).

Furthermore, the geometrical meaning of the symbols R and S is extended to cases where it is necessary to distinguish between octahedral and square planar structure or between cis and trans bis-(bidentate) octahedral metal chelates. For example, a bis-(bidentate) structure coded with R (or S) in front of the metal atom symbol:

(RM(.......) (.......))

is interpreted by the programme as a structure with cis-octahedral geometry and chirality corresponding to Δ (or Λ), whereas the same line formula without the chirality symbol:

(M(.......) (.......))

is interpreted as a trans-octahedral or square planar structure. The orientation of unsymmetrical bidentate chelates is implicit in the sequence of atomic symbols as coded in sidechains. In this way it is possible to specify unambiguously any geometrical isomer. For example, the five geometrical isomers of an octahedral M(AB)2X2 complex are coded as follows:

cis(X)-cis(A)-cis(B)	(RMX2(A....B) (A....B))
cis(X)-cis(A)-trans(B)	(RMX2(A....B) (B....A))
cis(X)-trans(A)-cis(B)	(RMX2(B....A) (A....B))
trans(X)-cis(A)	(MX2(A.....B) (A....B))

trans(X)-trans(A)   (MX2(A.....B) (B....A))

showing only one of the enantiomers (Δ) for each cis (X) isomers. In
a similar way facial and meridional isomers of a tris-(bidentate)
octahedral complex M(AB)3 can be distinguished:

fac-M(AB)3   (RM(A....B) (A....B) (A....B))

mer-M(AB)3   (RM(A....B) (A....B) (B....A))

as well as any other geometrical isomer of any octahedral or square
planar complex containing bidentate chelate rings.

From the examples shown above we can see that in most cases the
metal atom alone represents the principal chain and chelate ligands
the sidechains. The graph theoretical procedures outlined above are
also followed in the coding of formulae of coordination compounds.
All graphs representing the molecular structures of coordination
compounds (octahedral, tetrahedral or square planar) are planar and
convertible into spanning trees as exemplified in Figure 3.4. Graphs
of coordination compounds with multidentate chelate rings of any
complexity are treated likewise. Some examples are shown in Figure
3.5.

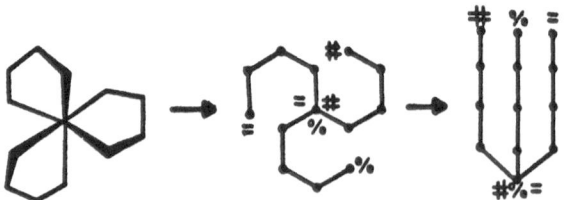

Figure 3.4 Construction of a spanning tree kenogram of an octahedral
tris(bidentate) complex corresponding to the line formula: (= %#M
(......#...) (......%...) (......=...))

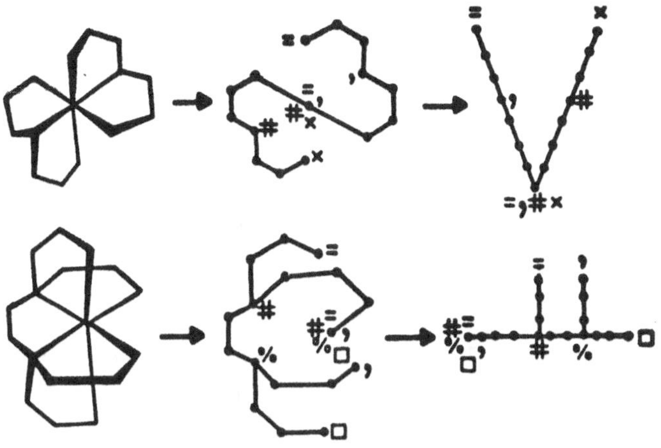

Figure 3.5 Further examples of spanning tree kenograms of coordination compounds: bis(tridentate) and hexadentate types.

3.1.5 Output from the programmes

The programmes BRACK and CODER produce, optionally, a detailed output of the topological analysis. This output should be used for checking purposes in the initial run on a new molecule. In routine runs it is suppressed.

3.1.5.1 Programme BRACK

BRACK first expands the line formula so that each chain atom and its sideatoms acquire six positions in a string of characters and blanks resolving at the same time the digits that specify the number of sideatoms. For example, the input formula of ethane, (CH3CH3), becomes (CHHH CHHH).

Furthermore, for each chain (pair of parantheses) BRACK forms an entry in a table showing the points of opening and closure of the chain, the total number of atoms and the number of chain atoms in that chain as well as before that chain.

3.1.5.2 Programme CODER

CODER performs an additional detailed analysis of the molecular topology.

The principal function of this programme is to find the correct arrangement in which the atoms are connected. The information is stored in an array of pairs of integers, each pair denoting a pair of bonded atoms. In addition, CODER forms a number of tables containing information about:

(1) Type numbers of all atoms. Each atom type is associated with a type number, which is used throughout the programmes for all non-numerical handling of atoms.

(2) Chain atoms, each of which is characterised by five integers specifying the code for atom type, position in the formula, degree, number of sideatoms, and types of sideatoms.

(3) Chelate rings, if there are any, with atom list numbers of ligating atoms. It is useful to have all these tables printed when a molecule is treated for the first time.

3.2 Lists of interactions

3.2.1 Programme MKLIST

MKLIST generates lists of all pairs of atoms forming bonds, triplets of atoms forming valence angles and quartets of atoms forming torsional angles, taking care that no interaction is counted twice. MKLIST also generates lists of pairs of atoms that are to be excluded from non-bonded calculations (1-2 interactions, or 1-2 and 1-3 interactions, depending on the choice of force field). An entry in one of those lists consists of a set of 2, 3 or 4 integers specifying atom indices of the atoms forming a bond, valence angle or torsional angle, a unique code specifying the type of interaction, and the so-called packed word (see below). Information about pairs of atoms to be excluded from non-bonded calculations is also coded as packed words. The packed words are stored on a temporary file.

3.2.2 Interaction codes

We have already mentioned that each of the 12 atom types has its own type number (Table 3.1), which identifies an atom of a particular type. In addition, the programmes utilise a 12*12 symmetric matrix (Table 3.2), in which any element $a(ij) = a(ji)$ has a unique integer value which identifies a pair of atom type numbers i and j.

Table 3.2. The pair interaction code matrix

		H	D	X	Q	O	A	K	C	N	M	T	P
		1	2	3	4	5	6	7	8	9	10	11	12
H	1	1											
D	2	2	3										
X	3	4	5	6									
Q	4	7	8	9	10								
O	5	11	12	13	14	15							
A	6	16	17	18	19	20	21						
K	7	22	23	24	25	26	27	28					
C	8	29	30	31	32	33	34	35	36				
N	9	37	38	39	40	41	42	43	44	45			
M	10	46	47	48	49	50	51	52	53	54	55		
T	11	56	57	58	59	60	61	62	63	64	65	66	
P	12	67	68	69	70	71	72	73	74	75	76	77	78

Composite codes are used for specification of valence and torsional angles: the pair code of the outer pair and the type number of the central atom for valence angles, and the pair codes of the outer and the inner pair for torsional angles. The interaction codes are :

bonds   a(ij)
valence angles   j * 100 + a(ik)
torsional angles   a(il) * 100 + a(jk)

with i,j,k and l being atom type numbers. Non-bonded exclusions are coded in a rather intricate way in nine-digit integer words, one set of words for each atom. The first digit counts the number of words in the set. The next two digits in the first word of the set denotes the atom unique to the set. The following characters of two digits in this and the subsequent words of the set denote the list numbers of all other atoms that are not allowed to interact in a non-bonded way with the unique one. No such pair of atoms is counted more than once.

3.2.3 Coding and decoding of integer words of interaction

In order to simplify the transfer of long lists of interactions between the programmes and the background memory we use a dense decimal packing scheme. It will pack up to five two-digit integers I1, I2, I3, I4 and I5 into a full word (INTEGER * 4) by successive multiplication by 100 and addition:

IWORD = (((((((I1*100)+I2)*100)+I3)*100)+I4)*100)+I5

or, equivalently:

IWORD=I1*100(N-1)+I2*100(N-2)+I3*100(N-3)+I4*100(N-4)+I5*100(N-5)

where N is the number of two-digit integers to be packed. This task is performed by the small subroutine ENCODE.

The i'th integer may be unpacked by a separate subroutine DECODE using the MOD function:

Ii = MOD(IWORD/100(i-1),100)

3.3 Molecular geometry

The next step is to obtain cartesian atomic coordinates. Depending
on the mode of operation, the programme can read coordinates from
cards or from a peripheral storage unit, or calculate them on basis
of the topological analysis.

3.3.1 Construction of molecular geometry

In Section 3.1.2 we have shown how to represent a molecular structure by a graph and how to reduce a graph containing circuits into a
spanning tree. The molecular tree concept has been found useful also
in developing a method for calculation of atomic coordinates from
bond lengths, valence angles and torsional angles.

The sequence of atomic symbols of chain atoms in a line formula can
be marked on a corresponding graph by adding arrows to the vertices
(Figure 3.6) thereby producing a directed graph. If a molecular
structure consists of a single chain (as polyglycine) or can be
reduced to one (as the spanning tree of a cycloalkane) the geometry
can be built up starting from one end of the chain and proceeding
atom by atom along the chain. A matrix method for doing this was
first put forward by Eyring (1932) who used coordinate transformations for the calculation of dipole moments of chain molecules.

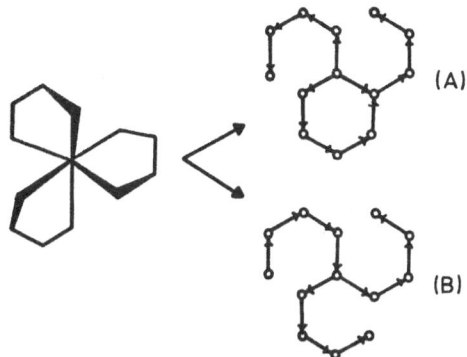

Figur 3.6 Two of the possible ways of producing a directed graph (kenogram is shown for simplicity) corresponding to a tris(biden-tate) coordination complex. (A) starts from the central metal atom and has three sidechains: (M(....)(....)(....)). (B) has only one sidechain: (.........M(....)....). The usage of the former representation is advisable when the molecular geometry has to be constructed by the programme.

In various studies on the conformations of polypeptide chains, a method of calculation of atomic coordinates that has been widely used involves the transformation of coordinates of a peptide unit, using rotation matrices and translation vectors, in either direction of the chain. Usually, a local coordinate system (LCS) is defined on each repeating unit and the coordinates are transformed the i-th to the (i-1)-th system, etc., until the first LCS is reached which is then taken as the global or reference coordinate system (RCS) (Nemethy and Scheraga 1965; Ramachandran and Sasisekharan 1968; Flory 1969; Tamburini et al. 1973). Structures derived from polycyclic molecules or those of branched molecules can be treated in a similar way by building up the structural fragments correspond-ing to the chains and sidechains as described above, and by properly assembling the fragments to the desired structure. Our programme REDUCE uses this approach and is capable of building the molecular geometry of a variety of structural forms.

<u>Table 3.3</u> Bond lengths between chain atoms

	O	A	K	C	N	M	T	P
O	1.48	1.41	1.36	1.426	1.36	1.90	1.50	1.58
A		1.097	1.32	1.47	1.24		1.50	
K			1.40	1.516	1.333		1.71	
C				1.541	1.472	1.80	1.82	1.84
N					1.451	2.00	1.67	1.49
M						2.51		
T							2.04	1.86
P								2.24

For atomic symbols see Table 3.1.

REDUCE contains a library of standard chain atom - chain atom bond lengths (Table 3.3). The data were taken mainly from Sutton (1965) and Gordon and Ford (1972). Some of the values adopted in the present version of the programme may not be the best choices, but for our purposes they are not objectionable since they serve merely to construct the initial geometry which may be far from the equilibrium one. In chapter 5 it will be shown that any such trial structure can be successfully minimised in a given force field.

Similarly, there is a library of valence angles which are assigned according to the hybridisation type of the chain atoms (Gordon and Ford 1972), and a collection of torsional angles (0, $\pi/3$, $\pi/2$, $2\pi/3$ and π). However, it is possible to specify any torsional angle through input cards (see Section 3.3.5) enabling the user to start calculations on any desired conformer.

We shall first consider the calculation of atomic coordinates of a simple unbranched chain molecule. The line formula for such a molecule will consist of a sequence of atomic symbols enclosed within a single pair of parantheses

$$(C \ldots\ldots C \ldots\ldots C \ldots\ldots \quad \ldots C \ldots\ldots)$$
$$\quad 1 \qquad 2 \qquad 3 \qquad\qquad n$$

where the C's stand for chain atoms, and it will correspond to a directed graph (a kenogram is shown):

$$\circ\!\longrightarrow\!\circ\!\longrightarrow\!\circ\!\longrightarrow\cdots\cdots\longrightarrow\!\circ$$
$$C \quad\; C \quad\; C \qquad\qquad\quad C$$
$$1 \quad\;\; 2 \quad\;\; 3 \qquad\qquad\quad\; n$$

The programme will assume standard bond lengths, taken from Table 3.3 according to atom types, and standard idealised valence angles according to the type of hybridisation of the chain atoms. Without additional information it will assume zero torsional angles around the chain bonds.

In the following we shall use three types of right-handed rectangular coordinate systems:

(1) A reference coordinate system (RCS), which is the global system for the molecule.

(2) A chain coordinate system (CCS), in which the coordinates of a sidechain will be defined.

(3) A local coordinate system (LCS) on each chain atom.

Let d(i) be the distance between chain atoms i and i+1, θ(i) the angle between chain bonds d(i-1) and d(i), and ϕ(i) the torsional angle defined by chain bonds d(i-1), d(i) and d(i+1) (Figure 3.7).

We choose an LCS on each chain atom i such that the Z(i) axis coincides with the bond d(i) (which we will call the emerging chain bond), and the Y(i) axis lies in the plane defined by the atoms forming 0 (pointing in the acute direction of θ(i)).

We define the transformation matrix, Eq. 3.1, and the translation vector, Eq. 3.2.

$$T_i = \begin{pmatrix} \cos\phi_i & \sin\phi_i & 0 \\ -\sin\phi_i \cos\theta'_i & \cos\phi_i \cos\theta'_i & \sin\theta'_i \\ \sin\phi_i \sin\theta'_i & -\cos\phi_i \sin\theta'_i & \cos\theta'_i \end{pmatrix} \qquad 3.1$$

$$\underset{\sim}{d}_i = \begin{pmatrix} 0 \\ 0 \\ d_i \end{pmatrix} \qquad 3.2$$

T(i) is the product of two matrices, T(i)(θ'(i)) and T(i)(ϕ(i)), representing rotation around the X axis by an angle θ'(i), the complementary to the bond angle θ(i) between chain bonds on atom i, and around the Z(i) axis by an angle ϕ(i). T(i) and $\underline{d}$(i) are used for the coordinate transformation from the i-th to the (i-1)-th LCS in the following way.

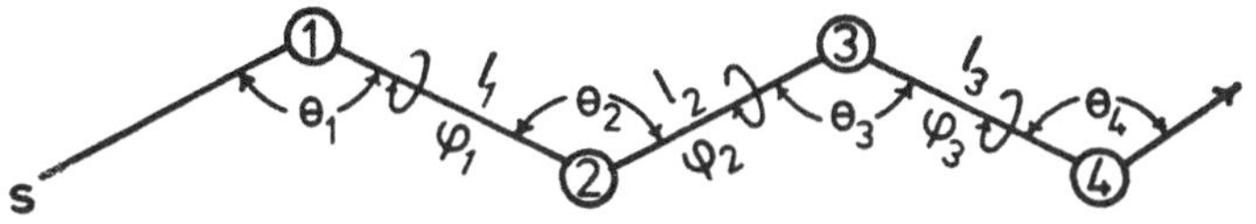

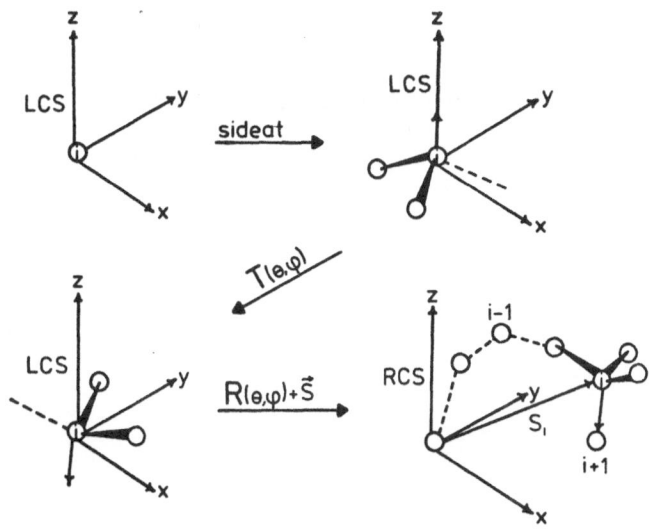

Figure 3.7 Portion of a chain showing the convention for labelling of chain atoms, chain bonds, valence and torsional angles. S = chain atom or sideatom for cyclic and acyclic structure, respectively.

Figure 3.8 Four steps in the building of a chain group.

We place the first chain atom (see Figure 3.8) in the origin of the LCS, and we add its sideatoms with the programme SIDEAT (Section 3.3.4). We rotate the coordinates of the whole group (in the LCS) using the matrix T(1) defined by $\theta(1)$ and $\phi(1)$:

$$\underline{r}^{(1)} = \underline{T}_1 \, \underline{r}'^{(1)} \qquad\qquad 3.3$$

where $\underline{r}^{(1)}$ is the set of coordinates of the sideatoms and $\underline{r}'^{(1)}$ the same coordinates in the standard orientation, and we put the resulting cluster into the RCS by the transformation

$$\underline{r}^{(1)}_{RCS} = R_1 \underline{r}^{(1)}_{LCS} + \underline{S}_1 \qquad\qquad 3.4$$

Since the RCS is coincident with the first LCS, matrix R in Eq. 3.4 is the unit matrix, and vector $\underline{S}(1)$ is the zero vector.

In general, however, the transformation that will put the i-th chain atom and its sideatoms into the RCS is given by Eq. 3.5:

$$\underline{r}^{(i)}_{RCS} = R_i \underline{r}^{(i)}_{LCS} + \underline{S}_i \qquad\qquad 3.5$$

where

$$R_i = \prod_{n=1}^{i-1} T_n \qquad\qquad 3.6$$

and

$$\underline{S}_i = \sum_{n=2}^{i} R_n \underline{d}_{n-1} = \sum_{n=2}^{i} \left(\prod_{m=1}^{n-1} T_m \right) \underline{d}_{n-1} \qquad\qquad 3.7$$

In this formulation R(i) is a matrix which accumulates the information about the partial coordinate transformations for each chain atom up to the atom i that is being treated. Likewise, $\underline{S}(i)$ is the end-to-end distance of a chain in the RCS prior to the transformation of the i-th group. The whole procedure is summarised in Figure 3.8, which shows how a chain atom (each chain atom in turn) is placed in the origin of its LCS in the standard orientation with the emerging chain bond along the +Z axis, the sideatoms are added, and the whole group is transformed in the LCS according to the corresponding values of θ and φ (Eq. 3.3) and, finally, from the LCS to the RCS (Eq. 3.5).

A molecular structure consisting of several chains is represented by a line formula having several pairs of parantheses which may be nested. REDUCE starts by building the sidechain enclosed within the innermost pair of parantheses. If there are more than one innermost sidechain of the same order, they are processed serially from left to right.

The coordinates of each of the sidechains, in the respective CCS, are computed using the transformations (3.3) and (3.5). They are transformed to the CCS of the higher level sidechain when the atom carrying the sidechain is put into place. Depending on the level of nesting of sidechains, sidechain coordinates are transformed from their original CCS either through one or more (higher level) CCS and eventually to the RCS, or directly to the RCS.

For example, the structure of a molecule having a line formula of the type

```
(....(....)....(....(....)....)....)
 a    b    b'   c    d    d'   c'   a'
```

will be treated by REDUCE in the following order: dd' in the CCS of dd'; cc', with dd' transformed to the CCS of cc'; bb' in the CCS of bb'; aa', with bb' and the complete fragment enclosed within cc' transformed to the CCS of aa'. The CCS of aa' is the RCS of the molecule.

3.3.2 Coordination compounds

The basic idea of transforming the completed molecular fragments, specified as sidechains, is applied in building the structures of coordination compounds.

At present, REDUCE is able to construct any octahedral or square planar complex with mono- and bidentate ligands.

Each chelate ring is treated as a sidechain and is constructed in the corresponding CCS so that the first ligating atom lies on the +Z axis and the rest of the chelate ring essentially in the +Y+Z plane. Thus for tris-(bidentate) and bis-(bidentate) structures the CCS of individual chelate rings are identical, and the ring coordinates of the existing chelate rings have to be transformed each time a new ring is put into place. The matrices used for chelate ring transformations are

$$
T_\Delta = \begin{pmatrix} 0 & -1 & 0 \\ 0 & 0 & -1 \\ 1 & 0 & 0 \end{pmatrix} \quad T_\Lambda = \begin{pmatrix} 0 & 1 & 0 \\ 0 & 0 & -1 \\ -1 & 0 & 0 \end{pmatrix} \quad T_t = \begin{pmatrix} -1 & 0 & 0 \\ 0 & -1 & 0 \\ 0 & 0 & -1 \end{pmatrix} \qquad 3.8
$$

From a single chelate ring with ligating atoms on the +Z and +Y axes matrices T_Δ and T_Λ will generate right-handed and left-handed helical distributions of chelate rings (with respect to a C(3) or a pseudo-C(3) axis). Their choice depends on the chirality specification on the central metal atom. For example, the chelate rings of a tris-(bidentate) structure coded with R specification will be transformed using T_Δ. If a formula for a tris-(bidentate) is coded with no chirality symbol on the metal atom, the T_t transformation will be used, resulting in two of the chelate rings being put into the same position trans to the third one. Similarly, bis-(bidentate) structures will be transformed using T_Δ, T_Λ or T_t resulting in Δ-cis, Λ-cis or trans configurations, depending on whether the formulae were coded with R, S or without chirality specification.

The present version of the programme is not able to build the geometry of a coordination compound with multidentate ligands. Addition of this facility would require a fairly elaborate set of rules for unambiguous coding of formulae. However, in many cases it is possible to use the programme repeatedly, each time building a segment of the complete structure consisting of one or more bidentate chelate rings, and then collate the output of each step into a proper sequence corresponding to the formula of the complete structure. The following examples illustrate the stepwise building of fac-bis-(diethylenetriamine) and (triethylenetetramine) -(ethylenediamine) structures from cis and trans bis-(ethylenediamine) structural fragments:

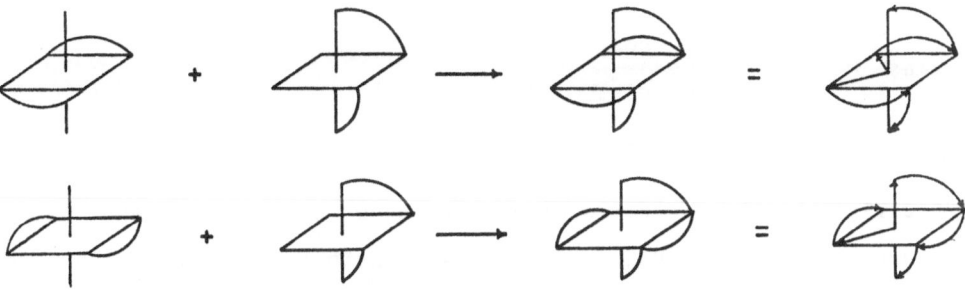

The resultant coordinates from each step can be assembled to a sequence corresponding to the overall formula with two sidechains of the form:

(M(......)(......)).

3.3.3 Incomplete structures

In a study of the conformations of molecules for which X-ray structures are known it is often advantageous to perform the energy
calculations and minimisations on the crystal structure conformations. For this purpose we use the programme CRYSTAL (see Section
2.2.6.2), which transforms fractional crystal coordinates into
cartesian atomic coordinates using the transformation matrix

$$T = \begin{pmatrix} A & B\cos\gamma & C\cos\beta \\ 0 & B\sin\gamma & C\sin\phi\sin\theta \\ 0 & 0 & C\cos\theta \end{pmatrix} \qquad 3.9$$

where A, B, C, α, β and γ are the unit cell dimensions, and ϕ and θ
are defined as

$$\cos\theta = \sqrt{1-\sin^2\theta}$$

$$\sin\theta = \cos\beta/\cos\gamma \qquad 3.10$$

$$\tan\phi = \frac{\cos\alpha}{\cos\beta \sin\gamma} - \frac{\cos\gamma}{\sin\gamma}$$

or, if $\beta= 90°$, $\phi= 90°$ and $\sin\theta= \cos\alpha/\sin\gamma$.

Usually only the positions of non-hydrogen atoms are reported by
crystallographers. However, it is possible to use a set of
non-hydrogen atom coordinates as an input and to have the structure completed by the programmes. This is done by the programme
HATOMS which adds hydrogens to primary, secondary and tertiary
carbon and nitrogen chain atoms assuming local tetrahedral geometry. Extension to other types of chain atoms and other geometries is possible and may be added at a later date.

HATOMS works as follows. For each chain atom (C or N) it finds the indices of the missing hydrogens in the array of cartesian coordinates, and it calculates the reference axis coincident with the sum of unit vectors pointing from that chain atom along its chain bonds (Figure 3.9).

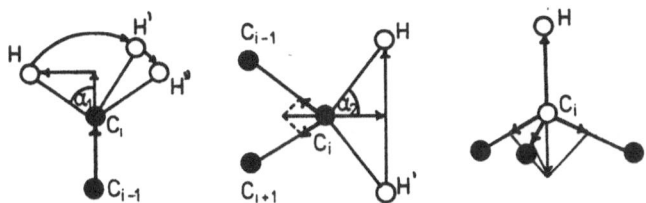

Figure 3.9 Adding hydrogens to a primary, secondary and tertiary carbon (or nitrogen) atom of approximate tetrahedral geometry.

Then the sideatom coordinates are assigned in reference to the Z-axis by the programme SIDEAT (Section 3.3.4) and transformed according to

$$r_{RCS}^{(i)} = M^{-1} r_{LCS}^{(i)} \qquad\qquad 3.11$$

where M is the rotation matrix that brings the chain atom reference axis into coincidence with the Z axis of the molecular coordinate system,

$$M = \begin{pmatrix} \cos\beta & \sin\beta\sin\alpha & \sin\beta\cos\alpha \\ 0 & \cos\alpha & -\sin\alpha \\ -\sin\beta & \cos\beta\sin\alpha & \cos\beta\cos\alpha \end{pmatrix} \qquad\qquad 3.12$$

and $r^{(i)}$ is the position vector of the i-th hydrogen atom. The angles α and β in Eq 3.12 are two of the Euler angles.

3.3.4 Sideatom positions

The programme SIDEAT performs the calculation of sideatom positions for a chain atom placed in the origin of its LCS. The necessary information which describes uniquely the geometry of the group is contained in three integer codes provided by the programme CODER for each chain atom. They are based on the input molecular formula and the special sideatom codes shown in Table 3.1.

Table 3.4   Sideatom Codes of Selected Atomic Groups

Group	Sideatom code
CH, NH, OH, KH, AH	1
CH_2, NH_2, KH_2, AH_2	2
CH_3, NH_3	3
K=Q	4
$K\langle^Q_H$, CX, NX	5
CXH, NXH, KHX, AHX	6
CXH_2, NXH_2	7
TQ_2	8

Sideatom codes of atomic groups are obtained as sums of sideatom codes of sideatoms (Table 3.1)

Chain atom-sideatom bond distances are selected according to the atom types of chain atoms.

KQH is distinguished from CX (or NX) on the basis of the total number of bonds (NUMBO in Table 3.7).

One of the codes used by SIDEAT specifies the type of the chain atom. The other specifies the number and types of the sideatoms and is derived by adding up the sideatom codes of the sideatoms (Table 3.4). The third code comes from the topological analysis of the input formula and represents the total number of bonds protruding from the chain atom (see Table 3.7). Two additional codes carry information about the chirality and about the position of the group in a chain.

The latter specifies whether the group, say CH3, is oriented so that the connecting chain bond is emerging (first CH3 group) or entering (last CH3 group) the chain atom.

Table 3.5 Sideatom bond lengths

	O	A	K	C	N	M	T	P
H	0.952	1.0	1.114	1.C9	0.98		1.33	1.44
D	0.952	1.0	1.114	1.C9	0.98		1.33	1.44
X	1.546		1.50	1.788	1.750	2.33		
Q			1.221			1.59	1.45	

For atomic symbols see Table 3.1.

Table 3.6 Vector components of sideatom

position vectors in chain atom coordinate systems

Vector	d	e	f
V_1	$a\,\sin\frac{\beta}{2}$	$-a\,\cos\frac{\beta}{2}\cos\frac{\gamma}{2}$	$-a\,\cos\frac{\beta}{2}\sin\frac{\gamma}{2}$
V_2	$b\,\sin\frac{\beta}{2}$	$-b\,\cos\frac{\beta}{2}\cos\frac{\gamma}{2}$	$-b\,\cos\frac{\beta}{2}\sin\frac{\gamma}{2}$
V_3	0	0	c
V_4	$a\,\cos\frac{\beta}{2}\sin\frac{\gamma}{2}$	$-a\,\sin\frac{\beta}{2}\sin\frac{\gamma}{2}$	$-a\,\cos\frac{\gamma}{2}$
V_5	$-b\,\cos\frac{\beta}{2}\sin\frac{\gamma}{2}$	$-b\,\sin\frac{\beta}{2}\sin\frac{\gamma}{2}$	$-b\,\cos\frac{\gamma}{2}$
V_6	0	$c\,\sin\frac{\gamma}{2}$	$-c\,\cos\frac{\gamma}{2}$
V_7	$-a\,\cos\frac{\gamma}{2}$	$-a\,\sin\frac{\beta}{2}$	$a\,\cos\frac{\beta}{2}$
V_8	$a\,\sin\frac{\gamma}{2}$	0	$-a\,\cos\frac{\gamma}{2}$
V_9	$-b\,\sin\frac{\gamma}{2}\sin\frac{\beta}{2}$	$b\,\sin\frac{\gamma}{2}\cos\frac{\beta}{2}$	$-b\,\cos\frac{\gamma}{2}$
V_{10}	$-c\,\sin\frac{\gamma}{2}\sin\frac{\beta}{2}$	$-c\,\sin\frac{\gamma}{2}\cos\frac{\beta}{2}$	$-c\,\cos\frac{\gamma}{2}$
V_{11}	$-a\,\sin\frac{\gamma}{2}$	0	$-a\,\cos\frac{\gamma}{2}$
V_{12}	0	$-a\,\sin\frac{\gamma}{2}$	$-a\,\cos\frac{\gamma}{2}$
V_{13}	0	0	b
V_{14}	0	0	$-a$

In the chain atom coordinate system, $(\underline{x},\ \underline{y},\ \underline{z})$ the position of the i-th sideatom is defined by the vector $\underline{V}_i = d\,\underline{x} + e\,\underline{y} + f\,\underline{z}$.

β is the valence angle between sideatoms.

γ is the complementary of the valence angle between chain atoms (where applicable).

a, b, c are the corresponding chain atom-sideatom bond lengths (see Table 3.5)

SIDEAT contains a library of standard sideatom - chain atom bond lengths (Table 3.5), the data being taken from Sutton (1965) and Gordon and Ford (1972), and a library of relevant angles specifying the idealised geometry on different chain atoms.

The present version of SIDEAT is able to produce the atomic coordinates of sideatoms of thirty different groups. Some of the most frequently encountered groups and their codes are shown in Table 3.4. The vectors defining the positions of sideatoms in the LCS are exemplified in Tables 3.6 and 3.7.

<u>Table 3.7</u> Usage of the sideatom position vectors
defined in Table 3.6

Group type	Position $^{x)}$	Vectors used	Examples
		Tetrahedral geometry (NUMBO=4)	
$-AB_3$	m,t	V_1 , V_2 , V_3	CH_3 , NH_3 , CX_3 , NX_3
	f	V_4 , V_5 , V_6	
	s	V_8 , V_9 , V_{10}	CH_3 , NH_3
$-AB_2 B'$	m,t	V_1 , V_2 , V_3	$CH_2 X$, $NH_2 X$, CHX_2 , NHX_2
	f	V_4 , V_5 , V_6	
$>AB_2$	m,t	V_1 , V_2	CH_2 , NH_2 , CX_2 , NX_2
	f	V_4 , V_5	
	s	V_8 , V_{11}	CH_2 , NH_2

each torsional angle. For example, a sequence of torsional angles of 180° will result in a fully extended chain structure, and a sequence of five torsional angles of 60°, with alternate signs, will give cyclohexane the chair conformation.

4 THE CONFORMATIONAL ENERGY AND ITS DERIVATIVES

Kjeld Rasmussen and Svetozar R. Niketić

4.1 Introduction

In this chapter we shall discuss the calculation of potential energy associated with molecular conformations. Having obtained cartesian atomic coordinates defining conformations, and lists of intramolecular interactions, as described in Chapter 3, we are ready to calculate a quantity which in the chemical literature is known as the total molecular potential energy or the conformational, steric, strain or intramolecular energy.

The conformational energy of a molecule can be expressed as a function V of all internal coordinates and interatomic distances, or as a function of atomic positions specified by some general coordinates. The function V is supposed to have local minima corresponding to the stable equilibrium conformations of a molecule in vacuo, neglecting intermolecular interactions.

The exact form of V is, of course, unknown. We assume that it can be suitably approximated by a sum of different types of energy contributions:

$$V = V_b + V_\theta + V_\phi + V_{nb} + V_e$$

The terms represent contributions to the total molecular potential energy V due to bond stretching and compression terms V_b, valence angle bending terms V_θ, internal rotational or torsional terms V_ϕ, non-bonded interactions V_{nb} and electrostatic or Coulomb interactions V_e. If there are other intramolecular mechanisms affecting

V, such as hydrogen bonding, corresponding terms may be added.

The total molecular potential energy V, as defined above, represents a measure of intramolecular strain of a molecule in vacuo in the hypothetical vibrationless state. The numerical value of V has no intrinsic physical meaning: its absolute value depends on the form of the potential functions and the choice of their parameters. However, differences in V for various conformations of the same molecule are related to molecular properties which can be measured experimentally. In addition, on the basis of the differences in V a relative energy scale can be established on which the positions on the energy scale of various known conformations can be illustrated and unknown conformations predicted.

We have no intention to give here a comprehensive account of energy functions that are being used in conformational analysis since a vast literature already exists on this subject. Instead, we shall focus our discussion on two essential aspects, the functional form of the energy terms and the calculation of V and its derivatives in internal and cartesian coordinates.

4.2 Intermolecular forces

It is convenient to classify molecular forces into intermolecular and intramolecular forces. Our use of potential energy functions for non-bonded intramolecular interactions is based on the theory of intermolecular forces.

4.2.1 Non-bonded interactions

Non-bonded interactions are the most significant of all the energy terms appearing in V, yet the model for their application embodies more simplifications than any other model energy function. The basic assumption is that we can use intermolecular potentials to treat quantitatively the intramolecular non-bonded interactions. Inter-molecular forces may as well be called interatomic, which is more adequate when discussing their application in conformational analysis, since they have been formulated to explain primarily the behaviour of monoatomic gases. This appears to be a good approximation although theoretically unjustifiable. Other assumptions comprise the pairwise additivity of non-bonded interactions, neglect of the intervening charge densities and of the directional dependency of the functions representing the chemical environment of the inter-acting atoms. The latter two effects are considered to be averaged out in the process of summation over all pairwise interactions (Schleyer et al. 1968).

The general form of the interatomic potential is shown in Figure 4.1. It is a sum of two terms representing two kinds of forces operating between two non-bonded atoms. At larger separations the net force is attractive and is due to the coupling of instantaneous dipoles induced in the interacting atoms. According to the theory of atractive forces developed by London (1937) the form of the potential function is given by

$$V(r) = -C/r^6$$

where the parameter C is a function of atomic polarisabilities and ionisation potentials of the interacting atoms (Pitzer 1959). Attractive forces are known as London dispersive forces or induced dipole - induced dipole interactions.

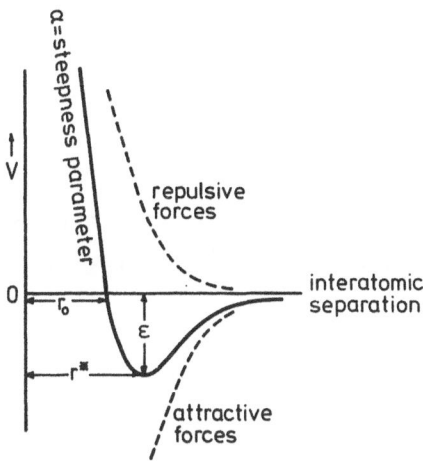

<u>Figure 4.1</u> General form of a non-bonded potential energy curve. For explanation of symbols see text.

At smaller separations where the electron clouds of two inter-acting atoms begin to overlap, the repulsive forces, known as overlap interactions, begin to dominate. They make the function very steep. The functional form of the repulsive or overlap interactions is approximated by either an exponential expression

V(r) = A exp (-Br)

or an inverse power expression

$$V(r) = A/r^n$$

where A and B are adjustable parameters.

The exponential form is theoretically more justified, since the wavefunction is exponential itself and, therefore, also the Coulomb and exchange integrals. However, the inverse power form is much used due to the empirical fact that it is just as effective in conformational analysis as the exponential.

Combining the potential functions for London dispersion interactions and overlap interactions, we arrive at the two most familiar expressions for interactions between two non-bonded atoms, both of which have been extensively used in conformational analysis:

$$V(r) = A \exp (-Br) - C/r^6$$

$$V(r) = A/r^n - C/r^6$$

The former is known as the modified Buckingham or exp-6 potential. The latter is a Lennard-Jones potential. It is mostly used with n = 12, though the choice of n = 9 and other values is also seen. In this form the formula is known as the Lennard-Jones 12-6 or 9-6 potential (Lennard-Jones 1931).

The Lennard-Jones potential can also be expressed in terms of the separation and the energy at the minimum of the potential function:

$$V(r) = \varepsilon(r^*/r)^{12} - 2\varepsilon(r^*/r)^6$$

$$V(r) = 2/3\, \varepsilon\, (r^*/r)^9 - \varepsilon(r^*/r)^6$$

The modified Buckingham potential can also be formulated in terms of ε and r^* and an additional parameter α, which governs the steepness of the exponential repulsion term, if we make the following substitutions:

$$A = e^{\alpha}(6\alpha/(\alpha-6))$$

$$B = \alpha/r^*$$

$$C = \alpha\varepsilon(r^*)^6/(\alpha-6)$$

In terms of α, ε and r^*, the modified Buckingham potential can be written as

$$V(r) = \frac{\varepsilon}{1-6/\alpha}\left\{6/\alpha\,\exp[\alpha(1-r/r^*)] - (r^*/r)^6\right\}$$

The Buckingham potential has the undesirable property of approaching $-\infty$ as the separation approaches zero. This may produce some difficulty in the case of an unrealistic initial conformation having very short separations between non-bonded atoms. The difficulty can be removed by assuming a hardsphere potential $V(r) = \infty$ (or rather a high but finite value) for $r \leqslant r(max)$, where $r(max)$ is the separation corresponding to the function maximum in the high-energy region.

The parameters r^* and ε in the two types of functions have the same physical meaning, which enables us to compare the functions more easily: r^* is the most stable interatomic distance in terms of potential energy of a pair of atoms and is observed, for example, as the nearest-neighbour distance in a crystal lattice of inert gases; ε, known also as the well depth, is the work required to separate two atoms from their most stable distance to infinity, and is related to the heat of sublimation of a crystal.

The parameters in the Lennard-Jones and modified Buckingham functions used in the theory of intermolecular interactions are adjusted to fit mainly four classes of experimental data: PVT data of gases (second virial coefficients); crystal properties (enthalphy of sublimation, nearest-neighbour distance); transport properties of gases (viscosity, diffusion, thermal conductivity); molecular scattering data.

Our programme uses either the modified Buckingham or a Lennard-Jones potential.

A usual practice in applying these potentials in conformational analysis is to cut off interaction between atoms at separations greater than about 5 A. We employ no cut-off because that may lead to erronous conclusions, as the longer-range attractive London terms are missing.

4.2.2 Electrostatic interactions

Interactions between partial charges on atoms can be treated as multipole expansions. Here we shall mention only the monopole approximation.

In the approximation of point charge - point charge interaction a simple Coulomb expression is used, giving the energy of interaction between two partial charges e(i) and e(j) separated by a distance r(ij) in a medium of dielectric constant D as

$$V(r) = e(i)e(j)/Dr(ij)$$

This term is summed over all pairs ij, again assuming pairwise additivity.

Point charges are either estimated so that they reproduce bond moments and molecular dipole moments, or they are obtained from molecular orbital calculations, as in the method proposed by Del Re (1958, 1963) for σ-bonded systems. It is difficult to estimate the effective dielectric constant; values between 3 and 5 have been used by various authors, mainly in conformational analysis of biopolymers, supported by the theoretical studies of Ramachandran and Srinivasan (1969). Also forms involving dependence of D on distance have been used; they are reviewed by Hopfinger (1973).

In our programme we adopt a value of D = 1 which is a valid approximation for the study of isolated molecules in vacuo. Or stated in a slightly different concept: D is embodied in the e's, which are treated as parameters.

4.2.3 Hydrogen bonding

Contributions from hydrogen bonding are very important in the conformational analysis of certain types of molecular systems, for instance biopolymers, and there have been many attempts to formulate and test potential functions which will describe this type of interaction (Ramachandran 1968). Among the most thorough studies of this problem is the work of Scheraga et al. (McGuire, Momany and Scheraga 1972) who have developed an empirical hydrogen bond potential function based on molecular orbital calculations.

Explicit treatment of hydrogen bonding is not included in the present version of our programme. It may be implemented relatively easily, but there are recent indications that this may not be necessary (Hagler et al. 1974). Proper treatment of hydrogen bonding requires sensible treatment of non-bonded interactions, and Coulomb terms are necessary. Already in 1954 Coulson pointed this out (Coulson and Danielsson 1954).

4.3 Intramolecular forces

In addition to interatomic interactions involving atoms that are not chemically bound to each other, there exist various other types of interaction. They can be classified into two-atom, three-atom and four-atom interactions, corresponding to deformations of bond lengths, valence angles and torsional angles. From another point of view they may be classified as bond interactions (bond compression and stretching, and bond torsion) and interactions between charge clouds of geminal bond regions (angle bending, Urey-Bradley). Under both concepts, various cross terms may be included.

4.3.1 Bond stretching

The most general form of the empirical potential function that has been used to represent a change in potential energy due to bond stretching and compression is

$$V(b) = K_1(b-b_o) + 1/2\ K_2(b-b_o)^2 + 1/6\ K_3(b-b_o)^3$$

Most conformational calculations use $K(1) = K(3) = 0$, giving harmonic potential functions. Attempts to include linear terms, which merely is another way of changing $b(o)$, were made by Altona and Sundaralingam (1970). Cubic terms were included by Allinger et al. (1971), and various crossterms (stretch-stretch, stretch-bend etc.) by Ermer and Lifson (1973).

An alternative form of the bond stretching potential function that takes care of the anharmonicity and the fact that dissociation occurs at finite energies is the Morse function (Morse 1929):

$$V(b) = D\exp[-2\alpha(b-b_o)]-2D\exp[-\alpha(b-b_o)] = D[(\exp[-\alpha(b-b_o)]-1)^2-1]$$

which includes the bond dissociation energy D and another parameter α, chosen so that the second derivative gives the stretching force constant.

Our programme can treat bond stretching potential with a quadratic function; a quadratic plus a linear or a cubic term or both; a Morse function; or an inverse power function. This latter possibility will be analysed elsewhere.

4.3.2 Bond torsion

Rotation around double bonds is so restricted that it corresponds to the breaking and making of bonds, that is, transformations between isomers rather than conformers. It must, therefore, be considered an intrinsic bond property, bond torsion.

Such a statement should not be made about rotation around single bonds. As evidenced by various spectroscopic, diffraction and thermodynamic measurements (Orville-Thomas 1974) the rotation around single bonds in polyatomic molecules is hindered by a potential energy barrier. Only a very minor fraction of this barrier is a property of a bond which to a fair approximation is a sigma bond. The barrier arises from the interactions between charge clouds on atoms in 1,4 positions to each other, and between charge clouds in the 1,2 and 3,4 bonding regions. Reviews as well as penetrating studies on the origin of torsional barriers are abundant. Suffice it to mention Wilson (1959), Sovers et al. (1968), Lowe (1969), Pethrik and Wyn-Jones (1969), Clementi and van Niessen (1971) and Orville--Thomas (1974).

The torsional potential can generally be expressed as a function of a torsional angle by a Fourier series, in which only the cosine terms are included since the function is even:

$$V(\phi) = 1/2 \sum_n K_n (1 + \cos kn\phi)$$

K(n) is the rotational or torsional barrier of the n'th term, and k is the multiplicity of the barrier.

Torsional potentials of threefold symmetric rotors, with both rotating groups having C(3v) symmetry (as H3C-CH3), or having C(3v) and C(1v) symmetries (as H3C-OH) are usually approximated by only a single term:

$$V(\phi) = K_3 (1 - \cos 3\phi)$$

Sixfold barriers originating from rotating groups having C(3v) and C(2v) symmetries (as H3C-NO2) are likewise represented by

$$V(\phi) = K_3 (1 - \cos 6\phi)$$

and are found to be very small (Birshtein and Ptitsyn 1966).

Due to the symmetry of the rotating groups (C(4v) and C(3v)), the torsional potential around a metal-ligand bond in octahedral metal ammine complexes is considered to be twelvefold. Intuitively, it is to be expected that such a barrier should be very low since there is very little difference in geometries between eclipsed and staggered conformations.

Torsional potentials around metal-ligand bonds have been formulated either with the term (Buckingham and Sargeson 1971)

$$V(\phi) = \frac{1}{2} K_{12} (1 + \cos 12\phi)$$

or with a sum of four terms which approaches zero for

$$K_1 = K_2 = K_3 = K_4$$

$$V(\phi) = \frac{1}{2} \sum_{n=1}^{4} K_n [1+\cos 3[\phi + (n-1)\pi/2]]$$

or they have been omitted from the calculations on the assumption that they are negligibly small according to the experimental evidence that the rotation of NH_3 groups in $[Co(NH3)6]3+$ is free (Kim 1960). Another evidence justifying the omission of the torsional contributions due to the metal-ligand bond rotations might be the experimental finding that the torsional barriers for bonds involving heavier atoms, Si-O, P-O or P-N, are very small in comparison with the C-C barriers of the same symmetry (Scott et al. 1961).

Torsional potentials around double bonds or bonds having partial double bound character may be similarly represented by the term

$$V(\phi) = 1/2 \ K(1-\cos 2\phi)$$

An alternative and perhaps more realistic way of formulating a torsional potential about a double bond is

$$V(\phi) = 1/2 \ K(\phi-\phi_o)^2$$

applicable in a range of small deviations of ϕ from equilibrium Positions.

For single bonds, as stated above, a torsional term should simply not appear. However, in order to be able to treat composite potential energy functions available in the literature, and to compare with other computations, we have included also the more traditional concept of intrinsic barriers for single bonds in our programme.

There are two ways in which an expression for the torsional energy can be implemented in a programme. We may consider only one torsional angle per single bond between chain atoms and compute its contribution to the total torsional potential. This is known as the group torsional model (Gleicher and Schleyer 1967). The other possibility is to consider all combinations of outer pairs of atoms, define the torsional angle for each, and sum over all the contributions. This is known as the bond torsional model (Wiberg 1965; Gleicher and Schleyer 1967), and is more appropriate in describing the torsional situation for a nonsymmetrical arrangement of groups. In this model the torsional barrier used to compute each individual interaction is taken approximately as the overall barrier divided by the number of contributions. Thus for a bond between two four-coordinated atoms, which takes part in nine bond torsional contributions, the barrier for each term is taken as one ninth of the barrier of the group torsional model. Both models can be used in our programme.

4.3.3 Angle bending

Geminal interactions are very difficult to parametrise into simple analytical forms. For the very reason of simplicity, these interactions are traditionally treated with terms of the same form as bond stretching:

$$V(\theta) = K_1(\theta-\theta_o) + 1/2\ K_2(\theta-\theta_o)^2 + 1/6\ K_3(\theta-\theta_o)^3$$

Cross terms of angle and torsion deformations have been used in some cases (Warshel and Lifson 1970).

4.3.4 Urey-Bradley potential

Another way to improve the force field is an attempt to introduce the expression known as the Urey-Bradley potential (Simanouti 1949). In the present context this represents a model which explicitly takes into account the geminal or 1,3-interactions (interactions between two atoms bound to a common atom). Geminal interactions are usually expressed by a quadratic plus a linear term in the non--bonded distance:

$$V(UB) = 1/2F(d-d_o)^2 + F'(d-d_o)$$

4.4 Force field parametrisation

4.4.1 Parameters and variables

Having reviewed the individual potential energy contributions we may now rewrite the expression for V in a more illuminating way (Figure 4.2).

$$
V = \left\{ \begin{array}{l} 1/2 \sum_{b} K_b (b-b_o)^2 \\[2ex] \sum D[\exp[-2\alpha(b-b_o)]-2\exp[-\alpha(b-b_o)]] \\[2ex] \sum (A/b^5 + B/b^9 + C/b) \end{array} \right\} \text{summed over all bonds}
$$

$$
+ 1/2 \sum_{\theta} K_\theta (\theta-\theta_o)^2 \qquad \text{summed over all valence angles}
$$

$$
+ \sum [1/2F(d-d_o)^2 + F'(d-d_o)] \text{ summed over all 1,3-interactions}
$$

$$
+ \left\{ \begin{array}{l} \sum 1/2 \sum_{n=1}^{N} K_n/N(1+\cos k\phi_i) \\[3ex] \sum 1/2 \ K(1+\cos k\phi) \end{array} \right\} \text{summed over all torsions}
$$

$$
+ \left\{ \begin{array}{l} \sum_{i>j} (A/r^9 - B/r^6 + e_i e_j/r) \\[4ex] \sum_{i>j} (Ae^{-Br} - C/r^6 + e_i e_j/r) \end{array} \right\} \begin{array}{l} \text{summed over all non-bonded} \\ \text{distances} \end{array}
$$

Figure 4.2 Molecular potential energy functions.

The quantities b, θ, d, φ and r are those variables: bond lengths, valence angles, 1,3-distances, torsional angles and non-bonded distances that characterise a given conformation. All other quantities are adjustable parameters that characterise the particular force field chosen for the study. Using this formula we may calculate the total molecular potential energy of any molecular conformation, in a given force field, which is specified as a set of functions with associated parameters.

It is appropriate at this point to add some general comments concerning the potential functions summarised in Figure 4.2. It has been repeatedly stressed that the functions used in computing the various contributions to the total molecular potential energy are empirical. This merely means that they were devised to be simple mathematical expressions, easy to handle both in theoretical considerations and in numerical applications. Nevertheless, they are qualitatively valid, and many of them are qualitatively well understood. Their capability to give a quantitative description of molecular systems is almost entirely dependent on the choice of the adjustable parameters. Looking at the collection of potential functions as exemplified in Figure 4.2 it becomes apparent that knowledge of a large number of adjustable parameters is required in order to compute V of a given molecule.

The problem of force field parametrisation has been mentioned many times in the chemical literature. Several good force fields have emerged from comprehensive calculations on large sets of molecules, together with thorough refinement of parameters based essentially on trial-and-error. For an example of a recent critical evaluation of some force fields see Engler, Andose and Schleyer (1973).

Force field parametrisation in the CFF approach has a different
objective. Firstly, the CFF method itself aims at refining
(optimising) the energy parameters. In other words, a good force
field becomes the ultimate result of CFF calculations. Therefore the
initial (trial) parametrisation becomes less critical although still
important since a good estimate of initial parameter values greatly
facilitates their refinement, from the points of view of numerical
stability of the optimisation process and of computer economy.

Secondly, in CFF calculations the parameters that enter the
potential energy functions are not to be identified with force
constants. There is an essential difference between force constants
which are characteristic of individual molecules and energy
parameters which are characteristic of a force field. Therefore we
speak about the parameters of the potential energy function for CH
stretching interaction in a given force field K(b) and b(o) on one
side, and about the CH stretching force constant for a particular CH
bond in a particular molecule on the other, this force constant
being derived in the course of the computations. This does not imply
any notions on the differences or similarities in numerical values
of energy parameters and the corresponding force constants.

Nevertheless, in practice spectroscopic force constants are used in
estimating trial values of energy parameters prior to CFF calculations.

4.4.2 Specification of energy functions and parameters

We shall now return to the programming system and show how a desired force field can be specified for conformational calculations. First of all, it is necessary to supply the programme with numerical values of all parameters needed in the computation of molecular potential energies of sets of molecules. The set of parameter values entered at the beginning of a computation defines the force field (FF) which is used for all molecules comprised in a single computation. Bookkeeping of energy parameters is done by the programme NPAR. Before we discuss it in more detail we shall mention the following additional possibilities to choose or modify a FF.

4.4.2.1 Global control parameters

The FF is specified with the global control parameters KJELD, NIKI and NBTYP. The present usage of these is given in Table 4.1, which is copied from our Input Manual. It is possible to specify whether to include all possible torsions around a single bond, or only one torsion involving the maximum number of chain atoms, through the setting of NIKI. The programme MKLIST will switch to one of the three possible modes in which the lists of torsional interactions are computed. Depending on the number of entries for torsional interactions involving a particular single bond, the torsional energy contribution will be computed according to either the group torsional concept or the bond torsional concept. Likewise, it is possible to choose treatment of 1,3-interactions either with a Urey-Bradley potential or with a non-bonded potential. Further, KJELD governs the choice of potential function for bond stretching (harmonic, Morse or inverse power). Finally, NBTYP chooses the potential function for non-bonded interactions (Lennard-Jones or Buckingham).

Table 4.1. Usage of parameters KJELD and NIKI

Subroutine	Parameter value	Performance
MKLIST	KJELD < 0	1,2 and 1,3 interactions are ex= cluded from the list of non-bond= ed interactions.
	KJELD ⩾ 0	1,2 interactions are excluded from the list of non-bonded in= teractions, whereas the 1,3 and higher interactions are included.
	NIKI < 0	All possible torsional angles (e. g. nine torsional angles for a single C-C bond) are included in the list of torsional interac= tions.
	NIKI = 0	One torsional angle per single bond is included in the list of torsional interactions.
	NIKI > 0	For metal-ligand bonds same as if NIKI = 0, for all other bonds same as if NIKI < 0.
MOLEC	KJELD < 0	Torsional and Urey-Bradley terms are included in the computation of total energy.
	KJELD ⩾ 0	Urey-Bradley terms are omitted.

	KJELD > 1	Torsional and Urey-Bradley terms are omitted.
EBOND and BFUNC	KJELD ⩽ 0	All bond contributions are calcu= lated with harmonic functions.
	KJELD = 1	All bond contributions are calcu= lated with Morse functions.
	KJELD > 1	All bond contributions are calcu= lated with inverse power func= tions.

The present assignment of global control parameters conforms to the authors' needs, but can easily be altered to meet the requirements of other users.

4.4.2.2 Function subroutines

For each type of energy contributions there is a subroutine that contains the corresponding expression for the potential energy func- tion (xFUNC subroutines, see later). A particular type of potential function can be changed simply by replacing the corresponding subroutine. In this way different users of the same basic programme have an unlimited choice to compose their own force fields.

4.4.2.3 Energy parameter input

NPAR is a fairly elaborate bookkeeping subroutine for treatment of energy parameters, which operates in one of four modes.

(1) It reads energy parameter cards (Table 4.2) containing the interaction type symbol (see below) and values of the parameters to be used in potential functions for the corresponding interactions. Parameters are counted, indexed, sorted into different arrays and written onto a direct access background memory.

(2) It fetches parameter values from a disk file and sorts them into arrays as mentioned above.

(3) It updates parameters in a disk file by values read from cards, and sorts them as above.

(4) It updates selected parameters after each cycle of parameter optimisation.

Each parameter card contains a symbolic representation of the interaction type and up to six parameter values (Table 4.2). Regular atomic symbols (see Table 3.1) are used to specify the interaction types: a string of two, three and four hyphenated symbols represent bonds, angles and torsions; two symbols joined by three hyphens a pair of non-bonded atoms; a symbol followed by two hyphens an atom participating in a non-bonded interaction; a symbol followed by a period a fractional point charge.

Arrays resulting from NPAR contain information on (1) the number of each type of interaction, (2) the codes in interactions together with numbers counting how many parameters are specified for each type of interaction, (3) the parameter values (Figure 4.3).

Table 4.2 Coding of energy parameters

(copied from Input Manual)

Type of interaction	Code	Energy parameters				
	COLUMNS: 1234567	COLUMNS: 11-20	21-30	31-40	41-50	51-60
Bond end	H-			(not implemented)		
Bond	N-H	K(b)	b(0)		(harmonic)	
		D(e)	a	b(e)	(Morse)	
		A	B	C	(inverse power)	
Valence angle	C-C-N	K(θ)	θ(0)	F	F'	d(0)
Torsional angle	H-N-C-C	K(φ)	k			
Atom	H--	A	B	e	(Lennard-Jones plus Coulomb)	
Atom pair	H---H	A	B	C	e (Buckingham or Lennard-Jones)	
Fractional charge	H.	e				

F, F' and d(0) = Urey-Bradley parameters

k = multiplicity of torsional barrier

Buckingham parameters must be specified for atom pairs

Units:

K(b) kcalmol(-1)Å(-2)   b(0) Å

D(e) kcalmol(-1)   a   Å(-1)   b(e) Å

A   B   C

K(θ) kcalmol(-1)rad(-2)   θ(0) rad

F   kcalmol(-1)Å(-2)   F'   kcalmol(-1)Å(-1)   d(0) Å

K (φ) kcalmol (-1)   k

A   (kcalmol (-1) A (9)) (1/2) B   (kcalmol (-1) A (6)) (1/2)

A   kcalmol (-1) * 10 (-4)   B   A (-1)

C   kcalmol (-1) A (6)   e   elementary charge

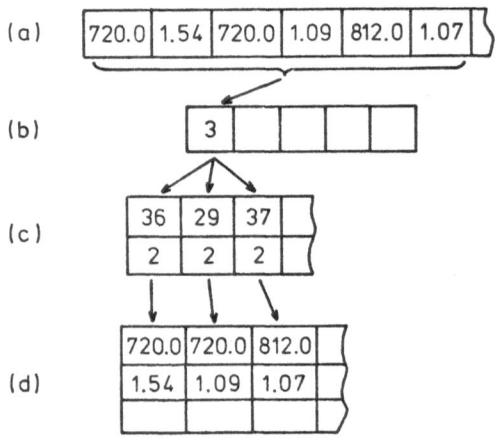

Figure 4.3 An example of the arrangement of data describing three bond stretching interactions in the arrays of programme NPAR. (a) Integral array of energy parameters; (b) Array showing the total number of interactions of bond stretching type; (c) For each bond stretching interaction a code number and the number of parameters are stored as pairs; (d) Parameter values are likewise stored in another array as pairs.

4.5 Energy calculations

4.5.1 Expansion of V in a Taylor series

The calculation of V of a given molecular conformation according to Figure 4.2 is a straightforward task assuming that we have at our disposal all the necessary potential energy functions and energy parameters, the lists of interactions and the cartesian atomic coordinates defining the conformation. This is about all we can get with Figure 4.2. Our interest in V is, however, much deeper. Primarily, we look for the sets of atomic coordinates, conformations, that correspond to the local minima of V and, therefore, we are interested in how V varies with changes in atomic coordinates.

If we suppose that the function V is well-behaved and that its first and second derivatives are continuous, which is generally the case, we may expand V in a Taylor series around the trial coordinates q(o):

$$V(q) = V(q_o) + \sum_{\alpha} (\frac{\partial V}{\partial q_\alpha})_o \delta q_\alpha + 1/2 \sum_{\alpha} \sum_{\beta} (\frac{\partial^2 V}{\partial q_\alpha \partial q_\beta})_o \delta q_\alpha \delta q_\beta + \ldots$$

This Taylor expansion forms the basis for the appliction of our energy minimisation methods which are described in the next chapter. The terms appearing on the right-hand side can be identified with some very important molecular properties, which enables us to correlate the calculations with experimental data, or, in the CFF concept, to embody the experimental data in the computations. In what follows we shall assume that the potential energy in the Taylor expansion is expressed in terms of cartesian atomic coordinates.

The first term, $V(q(o))$, represents the strain energy of a molecular conformation. It has already been pointed out that its numerical value depends on the choice of the force field, and therefore the difference in $V(q(o))$ for different conformations, rather than the values themselves, are physically meaningful and can be compared to conformational energy differences derived from experiments.

The second term contains the vector of first partial derivatives of the potential energy with respect to cartesian atomic coordinates. For a molecule in equilibrium the first derivatives of the energy are zero, and this term must vanish.

The third term contains the elements of the matrix of second partial derivatives of the potential energy with respect to cartesians, the Hessian matrix. At equilibrium the second derivatives of $V(q)$ are the force constants of the individual atomic displacements.

At present we do not consider the higher order terms of the Taylor series, assuming that they are small and that the possible improvement in our treatment would not justify the computational complexities involved in the calculation of third-order partial derivatives of the potential energy. One such extension has been described by Warshel (1971).

In the rest of this chapter we will describe the calculation of the total molecular potential energy and its first- and second-order partial derivatives, which will provide us with the necessary data for the energy minimisation procedures to be described in the next chapter.

4.5.2 Energy processing subprogrammes

At any point in the programming system where it is needed, calculation of the total molecular potential energy is performed by a
call of the subroutine MOLEC. This is merely a control programme for
a series of subroutines that calculate energy contributions from
bonds, valence angles, torsional angles, nonbonded and Urey-Bradley
interactions. (See Table 4.3). It calls successively the processing
subroutines xxxP (xxx = BOND, THETA, PHI, NBOND and UREY) and stores
the values of the corresponding energy contributions. Using the
global control parameters in MOLEC it is possible to omit the calculation of Urey-Bradley or torsional terms or both simply by pas-
sing over the calls of UREYP or PHIP subroutines. Furthermore, it is
possible to choose one of three modes of operation of MOLEC and its
subprogrammes, calculating either

(1) V only,
(2) V and $(\partial V/\partial x_\alpha)$, or
(3) V , $(\partial V/\partial x_\alpha)$ and $(\partial^2 V/\partial x_\alpha \partial x_\beta)$.

An independent option allows for calculation of the partial
derivatives of internals with respect to cartesians, $\partial s/\partial x$, which
are used in the vibrational analysis as the elements of the B matrix
and in the energy parameter optimisation.

Each of the processing subroutines xxxP runs over all of the
interactions listed by MKLIST, and for each entry it performs the
following steps:

(1) Decoding of the packed word.
(2) Calculation of interatomic distance
    (BONDP, NBONDP and UREYP) or angle (THETAP and PHIP).

Table 4.3 Subroutines used in the calculation of molecular potential energy classified according to the five types of energy
contributions

	Energy contributions				
	V_B	V_T	V_P	V_{NB}	V_{UB}
Processing subroutine calculates the corresponding internal coordinate or interatomic distance	BONDP	THETAP	PHIP	NBONDP	UREYP
Choice of parameters for potential energy functions	EBOND	ETHETA	EPHI	ENBOND	EUREY
Expressions for potential energy functions and derivatives	BFUNC	TFUNC	PFUNC	NBFUNC	UBFUNC

The distance between atoms i and j is calculated with the function
LENGTH

$$LENGTH(i,j) = \left\{ \sum_{\alpha=1}^{3} (x_{\alpha,i} - x_{\alpha,j})^2 \right\}^{1/2} ,$$

and the angle formed by atoms i, j and k is calculated as arc cos of the value of the function COSTHE

$$COSTHE(i,j,k) = \frac{\sum_{\alpha=1}^{3} (x_{\alpha,i} - x_{\alpha,j})(x_{\alpha,j} - x_{\alpha,k})}{LENGTH(i,j) \cdot LENGTH(j,k)}$$

(3) Next, xxxP calls the corresponding Exxx subroutine (xxx= BOND, THETA, PHI, NBOND and UREY; see Table 4.3), which chooses energy parameters from the arrays produced by NPAR according to the interaction type codes, and which in turn calls the corresponding xFUNC subroutine (x = B, T, P, NB and UB; see Table 4.3). The interaction type codes are calculated in Exxx from the atomic indices known from the Packed words and the atom codes. xFUNC calculates the energy contribution due to a given interaction as well as its first and second derivatives with respect to the corresponding internal coordinate or interatomic distance.

(4) After return to xxxP, the first and second derivatives of interatomic distances and valence or torsional angles with respect to cartesian coordinates are calculated with subroutines DIFBON (for distances) and DIFANG (for angles).

(5) Finally, the derivatives of the energy with respect to cartesians are formed and packed into the corresponding arrays.

This procedure is followed in the calculation of energy contributions of each of the five interaction types. It is summarised in Figure 4.4.

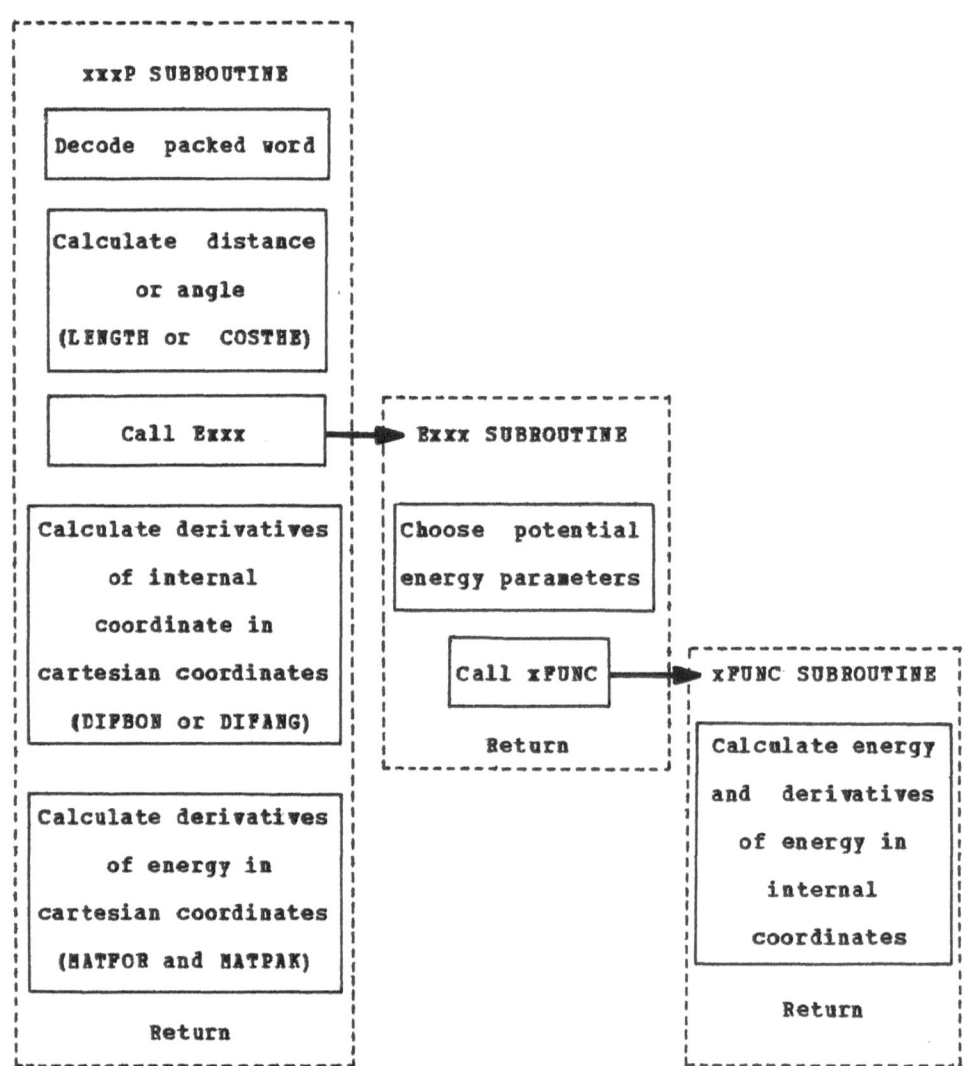

Figure 4.4 Programme sequence for the calculation of energy contribution and first- and second-order partial derivatives of energy with respect to cartesians.

It may be worthwhile pointing out that all the features described in this chapter concerning the implementation of the conformational energy calculation in our programming system exemplify the modular structure of the system. The calculation of energy and derivatives for each type of interaction is performed by an analogous set of subroutines. Therefore it is rather easy to add any other specific type of interaction by introducing a new set of subroutines written so that they parallel exactly any of the existing sets.

In the same way it is possible to extend the programme NPAR to enable it to treat any other type of energy parameter needed for any new type of potential energy contribution.

4.5.3 First- and second-order derivative calculation

The use of a Taylor expansion of $V(q)$ requires the evaluation of the first and second partial derivatives of potential energy with respect to cartesian coordinates. Both first and second derivatives are calculated analytically for all types of interaction. Using the chain rule we can express the partial derivatives of energy with respect to cartesians x in terms of energy derivatives with respect to internals r and derivatives of internals with respect to car-tesians:

$$\frac{\partial v}{\partial x_{\alpha,i}} = \frac{\partial v}{\partial r} \cdot \frac{\partial r}{\partial x_{\alpha,i}}$$

$$\frac{\partial^2 v}{\partial x_{\alpha,i} \partial x_{\beta,j}} = \frac{\partial^2 v}{\partial r^2} \cdot \frac{\partial r}{\partial x_{\alpha,i}} \cdot \frac{\partial r}{\partial x_{\beta,j}} + \frac{\partial v}{\partial r} \cdot \frac{\partial^2 r}{\partial x_{\alpha,i} \partial x_{\beta,j}}$$

The derivatives of energy with respect to internals and non-bonded distances for the potential energy functions included in the present version of the programmes are summarised in Table 4.4.

Table 4.4 Summary of the potential energy functions and their derivatives implemented in the conformational programmes

---

B o n d   s t r e t c h i n g

(1) Function with quadratic and cubic terms

$$V_b(b) = 1/2 K_b (b-b_o)^2 + 1/6 K'_b (b-b_o)^3$$

1-st derivative

$$V'_b(b) = K_b (b-b_o) + 1/2 K'_b (b-b_o)^2$$

2-nd derivative

$$V''_b(b) = K_b + K'_b (b-b_o)$$

(2) Inverse power function

$$V(b) = A/b + B/b^5 + C/b^9$$

1-st derivative

$$V'(b) = -A/b^2 - 5B/b^6 - 9C/b^{10}$$

2-nd derivative

$$V''(b) = 2A/b^3 + 30B/b^7 + 90C/b^{11}$$

(3) Morse function

$$V(b) = D[(e^{-\alpha(b-b_o)} - 1)^2 - 1]$$

1-st derivative

$$V'(b) = 2\alpha D[e^{-\alpha(b-b_o)} - e^{-2\alpha(b-b_o)}]$$

2-nd derivative

$$V''(b) = 2\alpha^2 D[2e^{-2\alpha(b-b_o)} + e^{-\alpha(b-b_o)}]$$

A n g l e   b e n d i n g

(4) Harmonic function

$$V(\theta) = 1/2 K_\theta (\theta-\theta_o)^2$$

1-st derivative

$$V'(\theta) = K_\theta (\theta-\theta_o)$$

2-nd derivative

$$V''(\theta) = K_\theta$$

(4a) Function with linear, quadratic and cubic terms

$$V(\theta) = K'_\theta (\theta-\theta_o) + 1/2 K_\theta (\theta-\theta_o)^2 + 1/3 K''_\theta (\theta-\theta_o)^3$$

1-st derivative $\qquad$ $V'(\theta) = K'_\theta + K_\theta (\theta - \theta_o) + 1/2 K''_\theta (\theta - \theta_o)^2$

2-nd derivative $\qquad$ $V''(\theta) = K_\theta + K''_\theta (\theta - \theta_o)$

T o r s i o n a l

(5) Cosine function $\qquad$ $V(\phi) = 1/2 K_\phi (1 + \cos k\phi)$

1-st derivative $\qquad$ $V'(\phi) = -1/2 K_\phi k \sin k\phi$

2-nd derivative $\qquad$ $V''(\phi) = -1/2 K_\phi k^2 \cos k\phi$

U r e y - B r a d l e y

(6) Function with linear
and quadratic terms $\qquad$ $V(r) = 1/2 F(r - r_o)^2 + F'(r - r_o)$

1-st derivative $\qquad$ $V'(r) = F'(r - r_o) + F'$

2-nd derivative $\qquad$ $V''(r) = F'$

N o n - b o n d e d

(7) Lennard-Jones function $\qquad$ $V(r) = A/r^9 - B/r^6 + E/r$

1-st derivative $\qquad$ $V'(r) = -9A/r^{10} + 6B/r^7 - E/r^2$

2-nd derivative $\qquad$ $V''(r) = 90A/r^{11} - 42B/r^8 + 2E/r^3$

(8) Buckingham function $\qquad$ $V(r) = A \exp(-Br) - C/r^6$

1-st derivative $\qquad$ $V'(r) = -AB \exp(-Br) + 6C/r^7$

2-nd derivative $\qquad$ $V''(r) = AB^2 \exp(-Br) - 42C/r^8$

There are three principal types of derivatives of internals with respect to cartesians, namely the derivatives of interatomic distances, valence angles and torsional angles. Their evaluation is based on the analytical formulae which can be obtained in a straightforward though tedious way. Without going into details of their derivation we shall present the formulae for the first and second derivatives of internals with respect to cartesians.

These formulae together with the formulae given in Table 4.4 are used in the two general equations for $\partial v/\partial x$ and $\partial^2 v/\partial x_\alpha \partial x_\beta$ given above. The first-order partial derivatives computed by the xxxP subroutines are stored in a one-dimensional array D. The second-order partial derivatives are computed by the subroutine MATPAK and are stored in a one-dimensional array DD. Since the matrix of second derivatives is symmetric, only the diagonal plus the upper triangle is stored, row-by-row.

4.5.3.1 Derivatives of interatomic distances

If we consider an interatomic distance b between atoms i and j with coordinates (x_{i1}, x_{i2}, x_{i3}) and (x_{j1}, x_{j2}, x_{j3}) we may define differences in cartesian components as

$$S_1 = x_{i1} - x_{j1}$$

$$S_2 = x_{i2} - x_{j2}$$

$$S_3 = x_{i3} - x_{j3},$$

and we may express the first partial derivatives of b with respect to cartesians through the derivatives of cartesian differences using the general relationship

$$\frac{\partial b}{\partial x_{\alpha\beta}} = \sum_{\gamma} \frac{\partial b}{\partial s_{\gamma}} \frac{\partial s_{\gamma}}{\partial x_{\alpha\beta}}$$

for $\alpha = i$, j; $\beta = 1$, 2, 3; and $\gamma = 1$, 2, 3; where

$$\frac{\partial b}{\partial s_{\gamma}} = \frac{s_{\gamma}}{b}$$

for $\gamma = 1$, 2, 3; and

$$\frac{\partial s_{\gamma}}{\partial x_{i,\beta}} = - \frac{\partial s_{\gamma}}{\partial x_{j,\beta}} = \delta_{\gamma\beta} \text{ (the Kronecker delta)}$$

Table 4.5. Derivatives of cartesian differences with respect to cartesians for interatomic distances

	s_1	s_2	s_3
$\partial/\partial x_{i,1}$	1		
$\partial/\partial x_{i,2}$		1	
$\partial/\partial x_{i,3}$			1
$\partial/\partial x_{j,1}$	-1		
$\partial/\partial x_{j,2}$		-1	
$\partial/\partial x_{j,3}$			-1

All second derivatives are zero.

Therefore, by substitution we get for the first derivatives of b with respect to cartesians:

$$\frac{\partial b}{\partial x_{\alpha,\beta}} = \pm \frac{s_\alpha}{b} = \pm \frac{x_{i,\alpha} - x_{j,\alpha}}{b}$$

The second partial derivatives of b with respect to cartesians can likewise be expressed as

$$\frac{\partial^2 b}{\partial x_{i,\alpha} \partial x_{j,\beta}} = \sum_{\gamma,\delta} \frac{\partial^2 b}{\partial s_\gamma \partial s_\delta} \cdot \frac{\partial s_\gamma}{\partial x_{i,\alpha}} \cdot \frac{\partial s_\delta}{\partial x_{j,\beta}} + \sum_\gamma \frac{\partial b}{\partial s_\gamma} \cdot \frac{\partial^2 s_\gamma}{\partial x_{i,\alpha} \partial x_{j,\beta}}$$

where

$$\frac{\partial^2 s_\gamma}{\partial x_{i,\alpha} \partial x_{j,\beta}} = 0$$

for all γ, leaving us with only the first term on the right-hand side, which can be further simplified to

$$\frac{\partial^2 b}{\partial s_\alpha \partial s_\beta} \cdot \frac{\partial s_\alpha}{\partial x_{i,\alpha}} \cdot \frac{\partial s_\beta}{\partial x_{j,\beta}} = - \frac{\partial^2 b}{\partial s_\alpha \partial s_\beta}$$

so that the formula for the second partial derivatives of b with respect to cartesians is reduced to

$$\frac{\partial^2 b}{\partial x_{i,\alpha}\,\partial x_{j,\beta}} = -\frac{\partial^2 b}{\partial s_\alpha\,\partial s_\beta} = \begin{cases} \dfrac{1}{b}\left\{1-\left(\dfrac{s_\alpha}{b}\right)^2\right\} \\[3em] \dfrac{s_\alpha s_\beta}{b^3} \end{cases}$$

where the upper expression applies for $\alpha=\beta$.

These equations are used in computations of first and second derivatives of interatomic distances with respect to cartesians, for interactions of bond length, non-bonded and Urey-Bradley types.

4.5.3.2 Derivatives of valence angles

Formulae for the first and second partial derivatives of valence angles with respect to cartesians are derived through the cartesian differences, in a manner analogous to that used above for interatomic distances:

$$\frac{\partial\theta}{\partial x_{\alpha,\beta}} = \sum_\gamma \frac{\partial\theta}{\partial s_\gamma}\cdot\frac{\partial s_\gamma}{\partial x_{\alpha,\beta}}$$

$$\frac{\partial^2\theta}{\partial x_{i,\alpha}\,\partial x_{j,\beta}} = \sum_{\gamma,\delta}\frac{\partial^2\theta}{\partial s_\gamma\,\partial s_\delta}\cdot\frac{\partial s_\gamma}{\partial x_{i,\alpha}}\cdot\frac{\partial s_\delta}{\partial x_{j,\beta}} + \sum_\gamma \frac{\partial\theta}{\partial s_\gamma}\cdot\frac{\partial^2 s_\gamma}{\partial x_{i,\alpha}\,\partial x_{j,\beta}}$$

For an angle defined by atoms i, j, and k the following cartesian differences are defined:

$$S_1 = x_{i,1} - x_{j,1}$$

$$S_4 = x_{k,1} - x_{j,1}$$

$$S_2 = x_{i,2} - x_{j,2}$$

$$S_5 = x_{k,2} - x_{j,2}$$

$$S_3 = x_{i,3} - x_{j,3}$$

$$S_6 = x_{k,3} - x_{j,3}$$

and the two bond lengths b_1 and b_2 :

$$b_1 = (S_1^2 + S_2^2 + S_3^2)^{1/2}$$

$$b_2 = (S_4^2 + S_5^2 + S_6^2)^{1/2}$$

The cosine of the angle θ is therefore defined as

$$\cos\theta = \frac{S_1 S_4 + S_2 S_5 + S_3 S_6}{b_1 b_2}$$

The first and second partial derivatives of θ with respect to cartesian differences are computed through the derivatives of cosθ using the following relationships:

$$\frac{\partial\theta}{\partial s_\alpha} = -\frac{1}{\sin\theta} \cdot \frac{\partial(\cos\theta)}{\partial s_\alpha} \qquad \text{for } \alpha = 1 \text{ to } 6$$

$$\frac{\partial^2\theta}{\partial s_\alpha \partial s_\beta} = -\frac{\cos\theta}{\sin^3\theta} \cdot \frac{\partial(\cos\theta)}{\partial s_\alpha} \cdot \frac{\partial(\cos\theta)}{\partial s_\beta} - \frac{1}{\sin\theta} \cdot \frac{\partial^2(\cos\theta)}{\partial s_\alpha \partial s_\beta}$$

The first derivatives of cosθ with respect to cartesian differences are

$$
\frac{\partial(\cos\theta)}{\partial s_\alpha} =
\begin{cases}
\dfrac{1}{b_1}\cdot\left(\dfrac{s_{\alpha+3}}{b_2} - \cos\theta\cdot\dfrac{s_\alpha}{b_1}\right) & \alpha = 1 \text{ to } 3 \\[4ex]
\dfrac{1}{b_2}\cdot\left(\dfrac{s_{\alpha-3}}{b_1} - \cos\theta\cdot\dfrac{s_\alpha}{b_2}\right) & \alpha = 4 \text{ to } 6
\end{cases}
$$

and the second derivatives of cosθ with respect to cartesian
differences are

$$
\frac{\partial^2(\cos\theta)}{\partial s_\alpha^2} =
\begin{cases}
-\dfrac{1}{b_1^2}\cdot\left(2s_\alpha\dfrac{\partial(\cos\theta)}{\partial s_\alpha} + \cos\theta\cdot\left(1-\dfrac{s_\alpha}{b_1}\right)^2\right) & \alpha = 1 \text{ to } 3 \\[4ex]
-\dfrac{1}{b_2^2}\cdot\left(2s_\alpha\dfrac{\partial(\cos\theta)}{\partial s_\alpha} + \cos\theta\cdot\left(1-\dfrac{s_\alpha}{b_2}\right)^2\right) & \alpha = 4 \text{ to } 6
\end{cases}
$$

$$
\frac{\partial^2(\cos\theta)}{\partial s_\alpha\, \partial s_{\alpha+3}} = -\frac{s_\alpha}{b_1^2}\cdot\frac{\partial(\cos\theta)}{\partial s_{\alpha+3}} - \frac{s_{\alpha+3}}{b_2^2}\cdot\frac{\partial(\cos\theta)}{\partial s_\alpha} -
$$

$$
-\frac{1}{b_1 b_2}\cdot\left(\frac{s_\alpha s_{\alpha+3}}{b_1 b_2}\cdot\cos\theta - 1\right) \qquad \alpha = 1 \text{ to } 3
$$

$$\frac{\partial^2 (\cos\theta)}{\partial s_\alpha \partial s_\beta} = \begin{cases} -\dfrac{1}{b_1^3 b_2}(S_{\alpha+3} S_\beta + S_\alpha S_{\beta+3}) + \dfrac{3}{b_1^4}(S_\alpha S_\beta \cos\theta) \\[2em] \qquad\qquad\qquad \alpha = 1 \text{ to } 3; \;\; \beta = 1 \text{ to } 3; \;\; \alpha \text{ not } = \beta \\[2em] -\dfrac{1}{b_2^3 b_1}(S_{\alpha-3} S_\beta + S_\alpha S_{\beta-3}) + \dfrac{3}{b_2^4}(S_\alpha S_\beta \cos\theta) \\[2em] \qquad\qquad\qquad \alpha = 4 \text{ to } 6; \;\; \beta = 4 \text{ to } 6; \;\; \alpha \text{ not } = \beta \end{cases}$$

$$\frac{\partial^2 (\cos\theta)}{\partial s_\alpha \partial s_\beta} = -\frac{S_{\alpha+3} S_\beta}{b_1^3 b_2} - \frac{S_\alpha S_{\beta-3}}{b_1^3 b_2} + \frac{S_\alpha S_\beta}{b_1^2 b_2^2}\cos\theta \quad \alpha = 1 \text{ to } 3; \;\; \beta = 4 \text{ to } 6$$

Derivatives of cartesian differences with respect to cartesians   are

$$\frac{\partial s_\alpha}{\partial x_{i,\beta}} = \frac{\partial s_{\alpha+3}}{\partial x_{k,\beta}} = -\frac{\partial s_\alpha}{\partial x_{j,\beta}} = -\frac{\partial s_{\alpha+3}}{\partial x_{j,\beta}} = \delta_{\alpha\beta} \qquad \alpha = 1 \text{ to } 3$$

$$\frac{\partial^2 s}{\partial x_\beta \partial x_\gamma} = 0 \qquad \text{for all } \alpha, \beta \text{ and } \gamma.$$

For convenience they are also tabulated in Table 4.6.

In the programmes, these formulae are computed step by step.

Table 4.6 Derivatives of cartesian differences with respect to cartesians for valence angles

	S_1	S_2	S_3	S_4	S_5	S_6
$\partial/\partial x_{i,1}$	1					
$\partial/\partial x_{i,2}$		1				
$\partial/\partial x_{i,3}$			1			
$\partial/\partial x_{j,1}$	-1			-1		
$\partial/\partial x_{j,2}$		-1			-1	
$\partial/\partial x_{j,3}$			-1			-1
$\partial/\partial x_{k,1}$				1		
$\partial/\partial x_{k,2}$					1	
$\partial/\partial x_{k,3}$						1

All second derivatives are zero.

4.5.4 Derivatives of torsional angles

Here again the general scheme outlined above for distances and valence angles is followed, but the derivation of the formulae is much more elaborate.

For a torsional angle defined by atoms i, j, k and l, cartesian differences can be defined as follows:

$$S_1 = x_{i,1} - x_{j,1} \qquad S_4 = x_{k,1} - x_{j,1} \qquad S_7 = -S_4 \qquad S_{10} = x_{l,1} - x_{k,1}$$

$$S_2 = x_{i,2} - x_{j,2} \qquad S_5 = x_{k,2} - x_{j,2} \qquad S_8 = -S_5 \qquad S_{11} = x_{l,2} - x_{k,2}$$

$$S_3 = x_{i,3} - x_{j,3} \qquad S_6 = x_{k,3} - x_{j,3} \qquad S_9 = -S_6 \qquad S_{12} = x_{l,3} - x_{k,3}$$

and the bond distances b_1, b_2, b_3 and b_4 can be defined in the usual way as

$$b_1 = (S_1^2 + S_2^2 + S_3^2)^{1/2}$$

$$b_2 = b_3 = (S_4^2 + S_5^2 + S_6^2)^{1/2} = (S_7^2 + S_8^2 + S_9^2)^{1/2}$$

$$b_4 = (S_{10}^2 + S_{11}^2 + S_{12}^2)^{1/2}$$

Vectors $\underline{b}_2$ and $\underline{b}_3$ correspond to the bond between the inner pair of atoms, $\underline{b}_2 = -\underline{b}_3$.

The torsional angle is computed with the following formula

$$\cos\phi = \frac{(\underline{b}_1 * \underline{b}_2) \cdot (\underline{b}_3 * \underline{b}_4)}{\|\underline{b}_1 * \underline{b}_2\| \cdot \|\underline{b}_3 * \underline{b}_4\|}$$

However, if we define the following new variables

$$T_1 = S_2 S_6 - S_3 S_5 \qquad\qquad T_4 = S_8 S_{12} - S_9 S_{11}$$

$$T_2 = S_3 S_4 - S_1 S_6 \qquad\qquad T_5 = S_9 S_{10} - S_7 S_{12}$$

$$T_3 = S_1 S_5 - S_2 S_4 \qquad\qquad T_6 = S_7 S_{11} - S_8 S_{10}$$

and

$$b'_1 = (T_1^2 + T_2^2 + T_3^2)^{1/2}$$

$$b'_2 = (T_4^2 + T_5^2 + T_6^2)^{1/2}$$

we can express $\cos\phi$ through the variables T (i = 1 to 6) and distances b' and b' analogously to what was done for $\cos\theta$:

$$\cos\phi = \frac{T_1 T_4 + T_2 T_5 + T_3 T_6}{b'_1 b'_2}$$

In this way it is possible to use the valence angle formulae developed above.

Derivatives of the variables T_i with respect to cartesians are given in Table 4.7.

Table 4.7 Derivatives of variables T (i=1 to 6) with respect to cartesians for torsional angles

	T_1	T_2	T_3	T_4	T_5	T_6
$\partial/\partial x_{i,1}$		$-S_6$	S_5			
$\partial/\partial x_{i,2}$	S_6		$-S_4$			
$\partial/\partial x_{i,3}$	$-S_5$	S_4				
$\partial/\partial x_{j,1}$		$S_6\ -S_3$	$S_2\ -S_5$		$-S_{12}$	S_{11}
$\partial/\partial x_{j,2}$	$S_3\ -S_6$		$S_4\ -S_1$	S_{12}		$-S_{10}$
$\partial/\partial x_{j,3}$	$S_5\ -S_2$	$S_1\ -S_4$		$-S_{11}$	S_{10}	
$\partial/\partial x_{k,1}$		S_3	$-S_2$	$S_{12}\ -S_9$	$S_8\ -S_{11}$	
$\partial/\partial x_{k,2}$	$-S_3$		S_1	$S_9\ -S_{12}$	$S_{10}\ -S_7$	
$\partial/\partial x_{k,3}$	S_2	$-S_1$		$S_{11}\ -S_8$	$S_7\ -S_{10}$	
$\partial/\partial x_{1,1}$					S_9	$-S_8$
$\partial/\partial x_{1,2}$				$-S_9$		S_7
$\partial/\partial x_{1,3}$				S_8	$-S_7$	

Out of 216 distinct second derivatives:

$$\frac{\partial^2 T_\alpha}{\partial x_{m,\beta} \, \partial x_{n,\gamma}}$$

72 are non-zero (two sets of 36 having the values +1 and -1, respectively).

In cases where $\sin\phi = 0$, that is where $\phi = k\pi$, we use the following equations:

$$\frac{\partial\phi}{\partial T_\alpha} = -\frac{1}{\cos\phi} \cdot \frac{\partial(\cos\phi)}{\partial T_\alpha}$$

$$\frac{\partial^2\phi}{\partial T_\alpha \partial T_\beta} = -\frac{1}{\cos\phi} \cdot \frac{\partial^2(\cos\phi)}{\partial T_\alpha \partial T_\beta}$$

$$\frac{\partial v}{\partial x_{i,\alpha}} = \frac{\partial^2 v}{\partial\phi^2} \cdot \frac{\partial\phi}{\partial x_{i,\alpha}}$$

$$\frac{\partial^2 v}{\partial x_{i,\alpha} \, \partial x_{j,\beta}} = \frac{\partial^2 v}{\partial\phi^2} \cdot \frac{\partial^2\phi}{\partial x_{i,\alpha} \, \partial x_{j,\beta}}$$

4.6 Numerical calculation of derivatives

Our programming system includes the possibility of numerical calculation of first and second partial derivatives of the total energy with respect to cartesians.

The programme TESTER calculates numerical first derivatives by Sterling's central difference quotient formula:

$$\frac{\partial v}{\partial x_i} = [V(\ldots x_i +h\ldots) - V(\ldots x_i -h\ldots)]/2h$$

Purely numerical calculation of second derivatives using the formulae

$$\frac{\partial^2 v}{\partial x_i^2} = [V(\ldots x_i +h\ldots) -2V(\ldots x_i \ldots) +V(\ldots x_i -h\ldots)]/h^2$$

$$\frac{\partial^2 v}{\partial x_i \partial x_j} = [V(\ldots x_i +h\ldots x_j +h) - V(\ldots x_i +h\ldots x_j -h\ldots)$$

$$- V(\ldots x_i -h\ldots x_j +h\ldots) + V(\ldots x_i -h\ldots x_j -h\ldots)]/4h^2$$

for diagonal and off-diagonal elements, respectively, would require too many function evaluations and would be intolerably time consuming. Therefore we calculate second derivatives by numerical differentiation of the analytically computed first derivatives:

$$\frac{\partial^2 v}{\partial x_i \partial x_j} - [V'(\ldots x_i +h\ldots x_j) - V'(\ldots x_i -h\ldots x_j)]/2h$$

where

$$v'_i = \frac{\partial v}{\partial x_i}$$

TESTER compares numerical derivatives with the corresponding analytical values and issues warnings when discrepancies greater than the preset value of 10(-9) are encountered. This option of the programming system is used on rare occasions when tests on new or modified energy processing subroutines are performed. Normally, the energy derivatives are computed analytically.

The value of the increment h in the formulae for numerical derivatives has been determined by trial-and-error comparison of derivatives for various points of Rosenbrock's two-variable function (Rosenbrock 1960):

$$f(x_1, x_2) = a(x_1^2 - x_2)^2 + (x_1 - b)^2$$

with the usual values a = 100, b = 1. Tests have been carried out for a range of h values between 10(-4) and 10(-9) both in single and double precision arithmetics. The most satisfactory results were obtained with h = 10(-6) in double precision. With this choice, the differences between numerical and analytical derivatives for Rosenbrock's function were always less than 10(-10).

5 ENERGY MINIMISATION

Svetozar R. Niketić and Kjeld Rasmussen

5.1 Statement of the problems

In this chapter we shall give a summary of the theory underlying the various methods for conformational energy minimisation. We shall focus our attention on the three methods which are implemented in our programming system: (1) the method of steepest descent; (2) the Davidon-Fletcher-Powell method; (3) the modified Newton method.

The general objective of energy minimisation in conformational analysis is to find a set of atomic coordinates defining a molecular conformation in such a way that this conformation corresponds to a minimum of the molecular potential energy in a given force field.

In all our problems we are dealing with polyatomic molecules. The molecular potential energy of such systems is a very complicated function of many independent variables. In general, it is a non-quadratic function, and, except in some particular cases, there are no constraints on the independent variables, either internal or cartesian coordinates. The problem of minimising such a function is known in mathematical language as the problem of unconstrained minimisation of a multivariable function. In mathematical literature various minimisation methods are often termed optimisation methods. Here we shall use the former term, reserving the latter for a different context. Recent mathematical literature abounds with studies of this problem, which, although it has a long history, has been made much more tractable with the development of fast and large computers. Among numerous studies we may mention the following comprehensive reviews describing recent developments and giving large bibliographies: Kowalik and Osborne (1968), Beveridge and Schechter (1970), Murray (1972), Brent (1973), Jakoby, Kowalik and Pizzi (1973), Altona and Faber (1973) and Gans (1976).

The general problem of minimising a multivariable function is still a very difficult one, and at present there exists no universally applicable and efficient algorithm. For this reason many types of minimisation methods have been developed for various applications. Even in a single area of application such as conformational analysis, it was found necessary to have access to several methods with different performances to meet the requirements of minimisation of different molecular conformations.

The analogy between the molecular features and their mathematical abstractions encountered in the theory of function minimisation can be illustrated as follows. As we have already pointed out, the molecular potential energy of a polyatomic molecule can be represented as a function of its conformation, which in turn is specified by a set of atomic coordinates. If we confine our discussion to cartesian atomic coordinates, the independent variables of the energy function of an N-atomic molecule will be the $3N$ cartesian coordinates. Any conformation, therefore, can be represented by a point in the $3N$-dimensional conformational space (hyperspace) or, alternatively, by a point on the potential energy surface (hypersurface) in $(3N+1)$-dimensional space. We may also express a given conformation as a vector whose components are the $3N$ cartesian atomic coordinates. Our task then becomes to find a point or points on the potential energy surface for which the function has a minimum or minima. This point (or points) will represent the equilibrium conformation(s) of a molecule.

In most practical applications of conformational energy minimisation the energy surface has proved to be very complicated so that the minima obtained using any of the current algorithms are generally local minima. The problem of finding the global minimum will be discussed later.

It is in general impossible to picture in the mind an energy hyper-surface and its characteristics. Only in the simplest cases where the problem can be reduced to two independent variables, can we visualise the conformational space and the potential energy surface. For example, we can express the potential energy of a regular helical polymer chain as a function of two independent variables, the torsional angles ϕ and ψ, and illustrate the energy contours on the familiar Ramachandran plots in the same way as on a topographical map we represent the altitude as a function of longitude and latitude.

5.2 Minimisation algorithms

All iterative methods for minimisation of a multivariable function consist of the following steps:

(1) Choice of the initial (trial) point, that is, selection of the starting conformation;

(2) Search strategy consisting of (a) exploratory movements through which we acquire information about the behaviour of the energy surface in the vicinity of the trial point; (b) choice of the direction of movement which can be either a predetermined direction or the direction of steepest descent or another downhill direction found on basis of the accumulated information about the energy surface; (c) choice of the distance of movement, the step length.

(3) Movement to a new point corresponding to a new conformation, hopefully with lower molecular potential energy.

(4) Termination criteria which will either allow the process to continue from step (2) or terminate the search within the required degree of accuracy or if the change in the function value gets smaller than a predetermined value.

Various minimisation methods available differ mainly in step (2), different approaches being used for the choice of direction and distance of movement. There are many different ways in which the minimisation methods can be classified. Since we confine our present discussion to methods which have found practical application in conformational analysis, we have found it convenient to classify them broadly into direct search methods and gradient or descent methods.

This classification does not include tabulation methods which are non-iterative methods for mapping of the chosen conformational subspace, usually one- or two-dimensional, over a mesh of coordinate values. These methods are widely used in the study of regular polymer conformations such as helical polypeptides, in which case the subspace consists of the torsional angles ϕ and ψ. The conformational energy surface of these systems is depicted as a contour diagram in the (ϕ, ψ)-plane, known as the Ramachandran steric map (Ramachandran et al. 1963, 1966; Ramakrishnan and Ramachandran 1965).

Examples of applications of various conformational energy minimisation methods are listed in Table 5.1, and, specifically for coordination compounds, in Table 5.2.

Table 5.1 Some applications of conformational energy minimisation methods

---

Tabulation methods

Mapping of conformational

subspace   a, b, c, d

Direct search methods

Rosenbrock's method   e, f

Modified SIMPLEX method   e

Pattern search   g, h, (w), bb

Descent methods

Steepest descent   e, i, j, k, l, m, n, o,

                                                          p, aa, (g), (w)

Parallel tangents (PARTAN)   g, q, r, (w)

Conjugate gradients   e, n

Smith's method of conjugate

directions   e

Powell's method of conjugate

directions   e, s

Davidon-Fletcher-Powell method   e, f, aa

Second-order gradient methods

(Newton methods)   f, t, u, v, w, x, y, aa, (g)

Miscellaneous

Non-simultaneous local

energy minimisation   z, (w)

---

a. G.N. Ramachandran, C.N. Venkatachalam and S. Krimm, Biophys.

    J. 6: 849 (1966).

b. J.B. Hendrickson, J. Am. Chem. Soc. 86: 4854 (1964).

c. A. Abe, P. Jernigan and P.J. Flory, ibid. 88: 631 (1966).

d. A.M. Liquori, A. Damiani and G. Elefante, J. Mol. Biol. 33: 439 (1968).

e. K.D. Gibson and H.A. Scheraga, Proc. N.A.S. 58: 420 (1967).

f. R.W. Busing, Programme WMIN, cit. from Acta Crystallogr. A28 (S4) Supplement (1972).

g. J.E. Williams, P.J. Stang and Schleyer, P.v.R., Ann. Rev. Phys. Chem. 19: 531 (1968).

h. E.M. Engler, J.D. Andose and P.v.R. Schleyer, J. Am. Chem. Soc. 95: 8005 (1973).

i. K.B. Wiberg, ibid. 87: 1070 (1965).

j. G.J. Gleicher and P.v.R. Schleyer, ibid. 89: 582 (1967).

k. N.L. Allinger, M.A. Miller, F.A. Van-Catledge and J.A. Hirsch, ibid. 89: 4345 (1967); and subsequent papers.

l. M. Bixon and S. Lifson, Tetrahedron 23: 769 (1967).

m. D.E. Williams, Science 147: 605 (1965).

n. I.D. Blackburne, R.P. Duke, R.A.Y. Jones, A.R. Katritzky and K.A.F. Record, JCS Perkin II 1973: 332.

o. P. DeSantis and A.M. Liquori, Biopolymers 10: 699 (1971).

p. J. Fournier and B. Waegell, Tetrahedron 26: 3195 (1970).

q. R.A. Scott and H.A. Scheraga, J. Chem. Phys. 44: 3054 (1966).

r. R.A. Scott, G. Vanderkooi, R. Tuttle, P. Shames and H.A. Scheraga, Proc. N.A.S. 58: 2204 (1967).

s. G.C.C. Niu, N. Go, and H.A. Scheraga, Macromolecules, 6: 91 (1973).

t. J. Jacob, H.B. Thompson and L.S. Bartell, J. Chem. Phys. 47: 3736 (1967).

u. A. Warshel and S. Lifson, J. Chem. Phys. 49: 5116 (1968).

v. B.H. Boyd, ibid. 49: 2574 (1968).

w. C. Altona and D.H. Faber, Fortschr. Chem. Forsch. 45: 1 (1974).

x. R.J. Ouellette, J. Am. Chem. Soc. 94: 7674 (1972) and Tetra-
    hedron 28: 2163 (1972).

y. W. Schubert and L. Schafer, J. Mol. Struct. 16: 403 (1973).

z. N.L. Allinger et al., J. Am. Chem. Soc. 93: 1637 (1971).

aa. K. Kildeby, S. Melberg and Kj. Rasmussen, Acta Chem. Scand.
    A31: 1 (1977).

bb. N.C. Cohen, Tetrahedron 27: 789 (1971).

Table 5.2 Application of force field calculations on coordination compounds.

Author(s)	Year	System	Approach	Ref.
Mathieu	1944	$[Co(l-pn)_2 X_2]^+$	Calc. of London dispersion forces	a
Corey & Bailar	1959	$[Co(en)_2 Cl_2]^+$ $[Co(en)_3]^{3+}$ Co(R-pn) ring	Calc. of non-bonded inter-actions (Note 1.)	b
Bagger	1964	$[Co(en)_3]^{3+}$ $[Co(2,3-bn)_3]^{3+}$ $[Co(tn)_3]^{3+}$	Calc. of non-bonded inter-action (Note 2.)	c

Buckingham et al.	1966	$[Co(en)_2(sar)]^{2+}$	Calc. of non-bonded inter-actions (Note 3.)	d
Woldbye et al.	1967	$[Co(tn)_3]^{3+}$	Calc. of non-bonded inter-actions (Note 2.)	e
Buckingham et al.	1967	$[Co(en)_2(aa)]^{2+}$ $[Co(Meen)_2(NO_2)_2]^{+}$ $[Co(trien)X_2]^{+}$ $[Co(NH_3)_4(Meen)]^{3+}$ $[Co(NH_3)_4(Sar)]^{2+}$	Calc. of non-bonded inter-actions (Note 3.)	f-i
Gollogly & Hawkins	1967	$[Co(R-mepenten)]^{3+}$	Mapping (Note 3.)	j
Buckingham et al.	1968	$[Co(Meen)_2(NO_2)_2]^{+}$	Calc. of non-bonded inter-actions	k
Gollogly & Hawkins	1969	en, pn, Meen chelate rings	Mapping	l

Gollogly & Hawkins	1970	$[Co(en)_3]^{3+}$ $[Co(en)_2X_2]^+$	Mapping	m
Rasmussen & Lifson	1970	$[Co(en)_3]^{3+}$ $[Cr(en)_3]^{3+}$	CFF (Note 4.)	n
Snow	1970	$\alpha\alpha$-$[Co(tetraen)Cl]^{2+}$	Boyd's pgm. (Note 5.)	o
Buckingham et al.	1970	β-$[Co(trien)(Pro)]^{2+}$	Boyd's pgm.	p
Gollogly, Hawkins & Beattie	1971	$[Co(en)_3]^{3+}$	Mapping	q
House et al.	1971	α-$[Co(en)(dtp)Cl]^{2+}$ β-$[Co(en)(dtp)Cl]^{2+}$	Boyd's pgm.	r
Geue & Snow	1971	$[Co(tn)_2CO_3]^+$ $[Co(tn)_3]^{3+}$ cis-$[Co(tn)_2(NO_3)_2]^+$	Boyd's pgm.	s

Gollogly & Hawkins	1972	tn chelate ring mono, bis and tris tn complexes	Mapping	t
Brubaker & Euler	1972	$[Co(picpn)(ox)]^{+}$ $[Co(picpn)Cl_2]^{+}$	Boyd's pgm.	u
Snow	1972	$\alpha\beta-[Co(tetraen)Cl]^{2+}$	Boyd's pgm.	v
Jurnak & Raymond	1972	$[Cr(tn)_3]^{3+}$	x-ray and Boyd's pgm.	w
Niketić & Woldbye	1973	$[Co(tn)_3]^{3+}$ $[Co(2,4-ptn)_3]^{3+}$	Wiberg's program (Note 6.)	x
DeHayes & Busch	1973	$[Co(en)X_4]^{-}$ $[Co(tn)X_4]^{-}$	Boyd's pgm.	y
DeHayes & Busch	1973	$[Co(dmdda)X_2]^{+}$	Boyd's pgm.	s
Dwyer, Geue & Snow	1973	$[Co(dien)_2]^{3+}$ $[Co(tn)_2CO_3]^{+}$	Boyd's pgm. symmetry-constrained	aa

Pratt & Ibers	1973	$P(C_6H_5)_3$	Busing's pgm. (Note 7.)	bb
Niketić & Woldbye	1973	$[Co(2,3-bn)_3]^{3+}$	Wiberg's pgm.	cc
Buckingham et al.	1974	$[Co(trien)(Gly)]^{2+}$	Boyd's pgm.	dd
Niketić et al.	1976	$[Co(tn)_3]^{3+}$ $[Co(2,4-ptn)_3]^{3+}$	CFF	ee
Niketić & Rasmussen	1977	$[Co(en)_3]^{3+}$ $[Co(2,3-bn)_3]^{3+}$	CFF	ff

Notes:

1. Rigid structure approximation. Mason and Kreevoy's (1955) non-bonded potential functions.

2. Rigid structure approximation. Simmons and Williams' (1964) non-bonded potential function. Molecular geometry calculated with a computer program.

3. Rigid structure approximation. Interatomic distances measured on Dreiding models. Non-bonded function of Hill (1948).

4. Consistent force field approach of Lifson and Warshel (1968). Minimisation performed by the method of steepest descent and modified Newton method.

5. Force field approach and Newton minimisation due to Boyd (1968).

6. Steepest descent minimisation and force field approach of Wiberg (1965).

7. Force field approach and minimisation program of Busing (1972).

References:

a.   J.-P. Mathieu, Ann. Phys. (Paris) 19: 335 (1944).

b.   E.J. Corey and J.C. Bailar, Jr., J. Am. Chem. Soc. 81: 2620 (1959).

c.   S. Bagger, Optisk Aktivitet og Konformationsanalyse i Koordinationskemi, Thesis, The Technical University of Denmark, 1964.

d.   D.A. Buckingham, S.F. Mason, A.M. Sargeson and K.R. Turnbull, Inorg. Chem. 5: 1649 (1966).

e.   F. Woldbye, Proc. Roy. Soc. A297: 79 (1967) and references to the work with A. Engberg, S. Bagger and G. Borch; this work was summarised in: F. Woldbye, Studier over Optisk Aktivitet, Polyteknisk Forlag, Copenhagen, 1969.

f.   D.A. Buckingham, L.G. Marzilli and A.M. Sargeson, J. Am. Chem. Soc. 89: 825 (1967).

g.   D.A. Buckingham, L.G. Marzilli and A.M. Sargeson, ibid. p. 3428.

h.   D.A. Buckingham, L.G. Marzilli and A.M. Sargeson, ibid. p. 5133.

i.   D.A. Buckingham, L.G. Marzilli and A.M. Sargeson, Inorg. Chem. 6: 1032 (1967).

j.   J.R. Gollogly and C.J. Hawkins, Austral. J. Chem. 20: 2395

(1967).

k. D.A. Buckingham, L.G. Marzilli and A.M. Sargeson, Inorg. Chem.
7: 915 (1968).

l. J.R. Gollogly and C.J. Hawkins, Inorg. Chem. 8: 1168 (1969).

m. J.R. Gollogly and C.J. Hawkins, ibid. 9: 576 (1970).

n. Kj. Rasmussen and S. Lifson, Unpublished work (1970); summarised in Kj. Rasmussen, Conformations and Vibrational Spectra
of Tris(diamine) Metal Complexes, Thesis, The Technical
University of Denmark, 1970.

o. M.R. Snow, J. Am. Chem. Soc. 92: 3610 (1970).

p. D.A. Buckingham, I.E. Maxwell, A.M. Sargeson and M.R. Snow,
ibid. 92: 3617 (1970).

q. J.R. Gollogly, C.J. Hawkins and J.R. Beattie, Inorg. Chem.
10: 317 (1971).

r. D.A. House, P.R. Ireland, I.E. Maxwell, and W.T. Robinson,
Inorg. Chim. Acta 5: 397 (1971).

s. R.J. Geue and M.R. Snow, J. Chem. Soc. (A) 1971: 2981.

t. J.R. Gollogly and C.J. Hawkins, Inorg. Chem. 11: 156 (1972);
The work of Gollogly and Hawkins has been summarised in:
C.J. Hawkins, Absolute Configuration of Metal Complexes,
Wiley, New York, 1971; and in: J.R. Gollogly, PhD Thesis,
University of Queensland, 1971.

u. G.R. Brubaker and R.A. Euler, Inorg. Chem. 11: 2357 (1972).

v. M.R. Snow, J. Chem. Soc. Dalton 1972: 1627.

w. F.A. Jurnak and K.N. Raymond, Inorg. Chem. 11: 3149 (1972).

x. S.R. Niketić and F. Woldbye, Acta Chem. Scand. 27: 621 (1973)
and ibid., 28: (1974).

y. L.J. DeHayes and D.H. Busch, Inorg. Chem. 12: 1505 (1973).

z. L.J. DeHayes and D.H. Busch, ibid. 12: 2010 (1973); summarised in: L.J. DeHayes, PhD Thesis, The Ohio State Universi-

ty, 1971.

aa. M. Dwyer, R.J. Geue and M.R. Snow, Inorg. Chem. 12: 2057 (1973).

bb. C. Pratt Brock and J.A. Ibers, Acta Cryst. B29: 2426 (1973).

cc. S.R. Niketić and F. Woldbye, Acta Chem. Scand. 27: 3811 (1973).

dd. D.A. Buckingham, P.J. Creswell, R.J. Dellaca, M. Dwyer, G.J. Gainsford, L.G. Marzilli, I.E. Maxwell, W.T. Robinson, A.M. Sargeson and K.R. Turnbull, J. Am. Chem. Soc. 96: 1713 (1974).

ee. S.R. Niketić, Kj. Rasmussen, F. Woldbye and S. Lifson, Acta Chem. Scand. A30: 485 (1976).

ff. S.R. Niketić and Kj. Rasmussen (to be published).

5.2.1 Direct search methods

All direct search methods are based only on function evaluation and comparison in some systematic way, and they do not require knowledge of any partial derivatives. They are relatively simple and easy to programme using a minimum of storage requirement. In particular cases they may be efficient, but, generally, they are very slow and poorly convergent.

The following direct search methods have been used in potential energy minimisation: the pattern search method of Hooke and Jeeves (1961), Rosenbrock's method (Rosenbrock 1960) and the modified simplex method of Nelder and Mead (1965). In most applications they have been abandoned in favour of the more powerful gradient methods. However, because of their simplicity they may be used in the preliminary stages of certain minimisation problems where crude but fast methods may be more economic.

5.2.2 Descent methods

The majority of current algorithms for multivariable function minimisation fall within this broad class. The oldest and simplest is the method of steepest descent (Cauchy 1847), and another well known one is Newton's method. Numerous other methods have been developed on the basis of these two. Descent methods differ from direct search methods in that they carry over information from one iteration to the following and use it to improve the search strategy. They usually involve computation of first order partial derivatives of the function in addition to the value of the function itself (first order gradient methods, conjugate gradient methods, variable metric methods and quasi-Newton methods); and sometimes also second order partial derivatives (second order or Newton methods). In addition, we include here those methods that do not involve actual computation of function derivatives, but which in all other respects behave as gradient methods (conjugate direction methods).

Descent methods that have found application in conformational analysis include: the method of steepest descent; the modified method of parallel tangents (PARTAN) originally due to Shah et al. (1964); the conjugate gradient method of Fletcher and Reeves (1964); Smith's method of conjugate directions (Smith 1962); Powell's method of conjugate directions (Powell 1965); and the Davidon-Fletcher-Powell method originally due to Davidon (1959) but reformulated by Fletcher and Powell (1963); and various second order gradient methods.

5.3 Unified approach to gradient algorithms

Before we proceed to describe the minimisation methods implemented in our programming system, we will introduce the notation and present a summary of the general form of the quadratically convergent gradient method from which most of the gradient algorithms can be derived. An algorithm is said to be quadratically convergent if it leads to a minimum within a finite number of steps or iterations when applied to a quadratic function. A quadratic function in n variables is any scalar function $f(x) = a + b'x + 1/2x'Ax$, where the scalar a, the n-vector b and the nxn symmetric matrix A are constants. All gradient methods except the method of steepest descent are quadratically convergent.

The problem can be formulated as follows. Find a local minimum of an unconstrained function $V(\underline{x})$ of n variables, which is assumed twice differentiable. At an arbitrary point k we shall denote the vector of independent variables $\underline{x}^k$:

$$\underline{x}^k = \left\{ x_1^k, x_2^k, x_2^k \ldots\ldots x_n^k \right\} ;$$

the gradient of $V(\underline{x})$, that is, the vector of first partial derivatives $\underline{g}$:

$$\underline{g}^k = \underline{g}(\underline{x}^k) = \left\{ \partial V/\partial x_1^k, \partial V/\partial x_2^k \ldots\ldots\partial V/\partial x_n^k \right\} ;$$

and the Hessian matrix, that is, the matrix of second partial derivatives G :

$$G_{ij}^k = G_{ij}(\underline{x}^k) = \partial^2 V/\partial x_i^k \partial x_j^k \qquad \text{for all i,j.}$$

Next, we shall expand the function $V(\underline{x})$ in a Taylor series around the minimum point $\underline{x}^o$ as follows:

$$V(\underline{x}) = V(\underline{x}^o + \underline{s}) \simeq V(\underline{x}^o) + \sum_i (\partial V/\partial x_i)^o s_i + \frac{1}{2}\sum_{ij} (\partial^2 V/\partial x_i \partial x_j)^o s_i s_j + R,$$

or, in matrix from, using Dirac bra-ket notation:

$$V(\underline{x}) = V(\underline{x}^o) + \langle g^o | s \rangle + \frac{1}{2}\langle s | G^o | s \rangle + R.$$

The remainder R becomes negligibly small when $\underline{x}$ is sufficiently close to the minimum $\underline{x}^o$; consequently the function $V(\underline{x})$ can be adequately approximated by a quadratic form in the vicinity of the minimum.

A minimum of a quadratic function

$$f(\underline{x}) = a + \langle b | x \rangle + \frac{1}{2}\langle x | A | x \rangle$$

can be found as follows. On differentiation, the necessary condition for minimum leads to a set of linear equations

$$\partial f(x)/\partial x = | b \rangle + A | x \rangle = 0,$$

having the solution

$$| x \rangle = -A^{-1} | b \rangle .$$

Since our equation arose from a second order Taylor expansion of a function $f(x)$, the resulting $| x \rangle$ will actually be the displacement of the point $\underline{x}$ from the minimum $\underline{x}^o$, which we will call $\underline{s}$:

$$| s \rangle = | x \rangle - | x^o \rangle = -A^{-1} | b \rangle.$$

In the case of a quadratic function, therefore, the minimum is reached in one step defined by $|s>$. Since the function $V(\underline{x})$ in general is not quadratic, its gradient vector and Hessian matrix at $\underline{x}^o$ are not known. Consequently we use an iterative technique, and $|s>$ only as an approximation to the direction in which the minimum is located, known as the search direction·

The basic algorithm for the iterative gradient minimisation can be represented as

$$|p^i> = H^i |g^i>$$

$$|s^i> = -|p^i>$$

$$|x^{i+1}> = |x^i> + |s^i> .$$

$|p^i>$ is the search direction. $|g^i>$ is the gradient evaluated at the i'th iteration, and H is a symmetric nxn matrix which characterises a particular algorithm·

$|s^i>$ is the step to be taken in the search direction $|p^i>$. The stepsize α^i is obtained by one-dimensional minimisation of $V(x^i - \alpha p^i)$ with respect to α^i, for instance by satisfying

$$\frac{d}{d\alpha_i} V(x^i - \alpha^i p^i) = 0.$$

$|x^{i+1}>$ is the displacement or correction of the current point at the i'th iteration. The displaced point $|x^{i+1}>$ is used as the starting point in the next iteration.

In the method of steepest descent the matrix H is the nxn identity matrix I, which remains fixed throughout the iterative process:

$$|p^i> = I|g^i> = |g^i>$$

In the Newton method the matrix H is the inverse of the Hessian matrix evaluated at the point x^i:

$$|p^i> = (G^i)^{-1}|g^i>$$

In general, the matrix H is a symmetric positive definite nxn matrix. A positive definite matrix is any symmetric matrix A (=A'), which gives $<x|A|x> > 0$ for any vector $|x>$. According to the way in which H is updated during minimisation, it characterises a particular variable metric algorithm.

The theoretical basis and the derivation of quadratically convergent variable metric methods can be found in the books mentioned earlier in this chapter and in the following papers: Myers (1966), Huang (1970), Huang and Levy (1970) and Adachi (1971).

5.4 Evaluation of minimisation methods

Mathematical literature abounds with critical and rigorous evaluations and comparisons of various minimisation methods (Fletcher 1965; Box 1966; Goldfeld et al. 1968; Fiacco and McCormick 1968; Pearson 1969; and any of the books cited at the beginning of this chapter). Most test functions used in these studies have relatively few independent variables. Although many methods have already been applied to minimisation of molecular potential energy (see Tables 5.1 and 5.2), comparisons of their performances and critical comments have been very sparse. Such a study might be useful from the application point of view since it is likely that the relative efficiencies of various minimisation methods might change with a marked increase in the number of independent variables, as in minimisation on medium and large molecules.

The first iterative minimisation algorithm to be applied in conformational analysis was the method of steepest descent, introduced by Wiberg in 1965. During the subsequent years various versions of the steepest descent method have been used. The first attempt to evaluate minimisation methods was done by Scheraga and coworkers (1967) who tested seven minimisation algorithms of direct search and first order gradient types, and concluded that the best of them in all respects was the variable metric method of Davidon (1959) and Fletcher and Powell (1963). It minimised successfully a function of more than 100 independent variables.

In a review from 1968, Schleyer and coworkers (Williams et al. 1965) claimed that their modification of the pattern search method of Hooke and Jeeves performed better than the method of steepest descent. They also noted that the method used by Jacob, Thompson and Bartell (1967) was superior to all the others but limited in use to quadratic energy functions.

More recently, Katritzky and coworkers (Blackburne et al. 1973) compared the method of steepest descent and the conjugate gradient method. They concluded that the methods lead to the same minima but that the conjugate gradient method exhibits better rate of convergence.

Finally, Altona and Faber (1973) reviewed five minimisation methods, again asserting the superiority of the second order gradient methods as applied in the programmes of Boyd (1968), Lifson and Warshel (1968) and Jacob, Thompson and Bartell (1967).

However, none of these studies presented any actual data which could form a basis for at least semiquantitative comparison of performances of minimisation algorithms applied in conformational analysis.

We have based our choice of minimisation techniques mainly on the experience of numerical mathematicians. In particular we have considered the following problems:

(1) The efficiency and the rate of convergence of different algorithms applied to a trial conformation which is far from minimum with respect either to the energy or to the geometry or to both.

(2) Optimisation of the time needed to compute the derivatives.

(3) Optimisation of the time needed to perform the linear search, the minimisation of the function along the search direction.

As is well known, there is no unique algorithm that can satisfy all efficiency requirements. Therefore, in our present version of the conformational programme we have adopted three minimisation methods of varying degree of complexity, all belonging to the class of gradient algorithms. The first of them is the method of steepest descent, which we often use in the preliminary stages of a minimisation. The third is the modified Newton method based on the algorithm of Gill, Murray and Picken (1972), which generally is used after the steepest descent. The second is the Davidon-Fletcher-
-Powell variable metric method, whose characteristics are intermediate between the two other methods, and which is best used alone.

5.4.1 The method of steepest descent

This method is seldomly used nowadays, particularly when main storage is not at a premium, since there are other gradient methods available which use essentially the same information, i.e. function values and first partial derivatives, but which exhibit much better convergence.

Nevertheless, after testing several similar procedures, we have found that a steepest descent method using a quadratic interpolation technique in the linear search is very efficient in minimising conformations which are far from minimum in energy or in both energy and geometry. Extremely distorted conformations may occur when initial coordinates are estimated by rough hand calculations or if cyclic structures are generated by the programmes without specification of torsional angles. Also a reasonable starting geometry when used with a poor force field will yield a highly distorted though artificial conformation, which must be minimised into a still artificial equilibrium before optimisation of the force field can take place. In such cases the steepest descent proved to be very expedient in bringing the conformation rapidly to its approximate region of minimum. The most remarkable performances were achieved in tests on highly distorted conformations of 40- to 60- atom molecules where much better minima, both in terms of energy and of geometry, were obtained from the same number of iterations, in comparison to the results of the Davidon-Fletcher-Powell method.

We have tested several steepest descent algorithms, differing in the line search technique.

(1) The steepest descent subroutine of Lifson and Warshel used one of three predetermined step lengths, 0.05, 0.002 and 0.0001, and performs the linear search by probing the function value in successive steps along the gradient, each time multiplying the step by 1.2 if it was successful, that is if V(new) < V(old), and by 0.5 if it was not. We abandoned the method because of its oscillatory behaviour for larger step lengths and its slowness for smaller step-lengths.

(2) The steepest descent algorithm of Rosen (1964) is similar to the above, but it includes the possibility to reverse the direction of linear search after unsuccessful steps. The method requires too many function calls per iteration and is poorly convergent.

(3) Wiberg's steepest descent programme (1965) is similar to (1) and (2) but does not allow for variation in steplength during line search. The method is stable but very slow.

(4) A new concept was tried. A series of five predetermined steplengths, chosen so that log(10) STEP = -3.0,-2.5, -2.0,-1.5 and -1.0, are applied each time a new search direction is found, and the point corresponding to the lowest function value is accepted for the next iteration. This extremely simple method performed better than any other on Rosenbrock's test function in two variables, but failed when applied to energy functions of big molecules.

(5) Our final choice was a steepest descent algorithm with the steplength calculated by quadratic interpolation. The line search is devised in such a way that a minimum is eventually bracketed by three equidistant points, α°, α^{k-1} and α^{k}, which are used to fit the parabola and compute its minimum analytically in the following way.

If we set the initial point on the search line to zero, the three points that bracket the minimum will be 0, α and 2α. The corresponding function values are V1, V2 and V3. The coefficients in the quadratic form through these three points are obtained by solving the set of linear equations

$$Q|a> = |v>$$

where the matrix Q is

$$Q = \begin{pmatrix} 0 & 0 & 1 \\ \alpha^2 & \alpha & 1 \\ 4\alpha^2 & 2\alpha & 1 \end{pmatrix}$$

For the coefficient vector |a> we get

$$a1 = (V1 - 2V2 + V3)/2\alpha^2$$

$$a2 = (4V2 - V3 - 3V1)/2\alpha$$

$$a3 = V1$$

Substituting these into a quadratic form, differentiating with respect to α, and setting $dV/d\alpha = 0$, we find the minimum as

$$\alpha_{min} = \frac{\alpha}{2} \cdot \frac{3V1 - 4V2 + V3}{V1 - 2V2 + V3}$$

The optimal initial steplength was estimated by trial-and-error and fixed at 0.001.

In order to make the algorithm function equally well for problems of widely differing dimensions (2-200), the search direction is normalised:

$$|p^k> = -|g^k>/\|g^k\|$$

Minimisation is terminated if any of two criteria is fulfilled:

$$V(x^k) - V(x^{k+1}) < \varepsilon 1$$

$$\|s^k\| < \varepsilon 2 \qquad\qquad \varepsilon 1 = \varepsilon 2 = 10(-6)$$

In the use of the method of steepest descent, termination criteria are of minor importance, since it slows down considerably when $|g^k>$ becomes small close to the minimum. Therefore we usually impose a limit of between 10 and 50 iterations, and the switch to another method, usually then modified Newton.

5.4.2 The Davidon-Fletcher-Powell method

Our next choice was one of the variable metric methods: the Davidon--Fletcher-Powell method (DFP) originally due to Davidon (1959) and extended by Fletcher and Powell (1963). It is generally agreed that DFP is the best general purpose minimisation procedure, but until recently it was only scarcely used in conformational analysis.

The central feature of the algorithm is the matrix H(i) which is used in computing the descent direction. The procedure starts with an arbitrary matrix H(o) which is updated after each iteration according to the information accumulated from previous iterations: the former H matrix, the difference between gradients and the steplength vector. Updating of H is done with the recursion formula

$$H^{i+1} = H^i + \frac{|dx><dx|}{<dx|dg>} - \frac{H^i|dg><dg|H^i}{<dg|H^i|dg>}$$

where $|dg> = |g^{i+1}> - |g^i>$ and $|dx> = |x^{i+1}> - |x^i>$.

The process of modifying the H matrix eventually leads to the inverse of the Hessian matrix. The recursion formula used here is only one of several described in the literature (Huang 1970; Huang and Levy 1970). There are other formulae which yield a null matrix at convergence.

Since the DFP is thoroughly described in the literature (see Jacoby, Kowalik and Pizzo, 1973, pp. 137-150), we shall give here only a brief summary of the salient points of its implementation in our conformational programme.

(1) Starting conditions. We follow the usual practice and set the initial matrix H(o) equal to the identity matrix. This means that the first DFP iteration is identical to the steepest descent iteration.

(2) Stopping conditions. Minimisation is terminated if a predetermined number of iterations is exceeded and also if the H matrix ceases to be positive definite. Usually the minimisation is stopped only after n iterations have been performed where n is the number of independent variables. For our purposes, this is an unneccessarily large number. We therefore employ the two termination criteria

$$\|p^i\| < \varepsilon 2 \text{ and } \|s^i\| < \varepsilon 2 \quad , \quad \varepsilon 2 = 10(-8)$$

(3) Restarting conditions. There are two ways in which the programme can be restarted. At the i'th iteration we can set H(i) = H(o) or = H(i-1). Restart in the steepest descent direction is performed if the minimum can not be bracketed by the linear search, or if the current steplength becomes smaller than a prescribed number ($\| s(i) \|$ <ϵ_2). Restart with the penultimate H matrix is done if there is evidence that the recursion formula will be numerically unstable due to zero denominators:

$$<dx|dg> < \epsilon_1 \text{ and } <dg|H^i|dg> < \epsilon_1 \quad , \quad \epsilon_1 = 10(-16)$$

(4) Linear search. Computational experience (Huang and Levy 1970, and others) indicates that the linear search should be rather precise. Therefore DFP minimisations are usually coupled with a line search based on a quadratic or cubic interpolation. We use a cubic interpolation derived from the function values V1 and V2 and the first derivatives g1 and g2 computed at two points x1 and x2. If we denote the distance between the two points we can set up a system of four equations, the cubic equation and its first derivative in the two points. The solution of these equations gives us the values of the coefficients of the third order polynomial ax + bx + cx + d = 0:

$$a = \frac{2(V1 - V2)}{\alpha^3} + \frac{(g1 - g2)}{\alpha^2}$$

$$b = \frac{3(V2 - V1)}{\alpha^2} - \frac{(2g1 - g2)}{\alpha}$$

$$c = g1$$

$$d = V1$$

On differentiation of the polynomial and setting the first derivative equal to zero we get a quadratic expression which gives the minimum:

$$x_{min} = \alpha \cdot \left(1 - \frac{g2 - z + w}{g2 - g1 + 2w}\right) = \alpha \cdot \left(1 - \frac{g2 + z - w}{g2 + g1 + 2z}\right)$$

where

$$z = g1 + g2 - \frac{3(V2 - V1)}{\alpha}$$

and

$$w = \pm (z^2 - g1 g2)^{\frac{1}{2}}.$$

Finally, we may mention, without going into computational details, that the linear search operates such that it ensures the bracketing of the minimum by two points. Therefore, the formula for x_{min} is applied only if $g1 < 10(-16)$ and $g2 > 10(-16)$. Another safeguard is a check on the value of the denominator in that formula: if it gets close to $10(-16)$, a quadratic interpolation is used instead:

$$x_{min} = \alpha \cdot \frac{g1}{g1 - g2} = \alpha^2 \cdot \frac{g1}{V1 - V2 + g1\alpha}$$

5.4.3 The modified Newton method

Our final choice was a second order gradient method which was adapted from the modified Newton algorithm MNA developed by Gill, Murray and Picken (1972).

The basic iteration steps of any Newton method follow directly from differentiation of a second-order Taylor expansion of the function $V(\underline{x})$:

$$V(\underline{x}) = V(\underline{x}^k) + \langle g^k | \Delta x^k \rangle + - \langle \Delta x^k | G^k | \Delta x^k \rangle$$

$$|g(x)\rangle = |g^k\rangle + G^k |\Delta x^k\rangle = 0$$

$$|\Delta x^k\rangle = - (G^k)^{-1} |g^k\rangle$$

where

$$|\Delta x^k\rangle = |x^{k+1}\rangle - |x^k\rangle = |s^k\rangle.$$

The search direction is defined by the gradient vector and by the Hessian matrix, and it is obtained by solving the set of linear equations

$$G^k |p^k\rangle = - |g^k\rangle.$$

In contrast to the other gradient methods, the Newton method requires more computer work since it involves calculation of second-order derivatives and solution of a set of linear equations in each iteration.

A positive definite Hessian matrix ensures the descent direction of $|p^k\rangle$. However, it may happen that the Hessian becomes singular or numerically not sufficiently positive definite. In such cases it is not possible to apply the Newton method directly; consequently various methods have been proposed (Jakoby, Kowalik and Pizzo 1973; Murray 1972) to overcome this problem. Most of them use some kind of modification of the Hessian (Goldfeld et al. 1966, 1968; Marquardt 1963; Fiacco and McCormick 1968; Matthews and Davies 1971) whereby they form another matrix which is positive definite and which is used instead of the original Hessian for the calculation of the descent direction.

A modification of the Newton method which is very fast, and yet reliable, is the MNA developed by Murray and co-workers (Gill, Murray and Picken 1972[2]; Murray 1972, pp. 64-69). The essence of the method is the Cholesky factorisation of the Hessian into lower triangular, diagonal and upper triangular matrices:

$$G^k = L^k D^k L'^k$$

The factorisation is possible if G is positive definite. If, during factorisation, there is indication that G is not positive definite, it is modified simultaneously so that the resulting decomposition corresponds to some other matrix G, which would have been obtained from G by the addition of a diagonal matrix E:

$$L^k D^k L'^k = G^{-k} = G^k + E^k$$

The theoretical basis of the method is described in detail by Gill, Murray and Picken (1972) and by Murray (1972, pp. 64-69). The latter reference also contains a critique of a number of modified Newton methods, pointing out their numerical instability.

In the MNA method, the descent direction $|p^k>$ is computed in two ways. If at the k'th iteration the norm of the gradient vector is greater than zero (in practice, if $\|g^k\| > \epsilon 2$), the system of linear equations is solved for $|p^k>$:

$$L^k D^k L'^k |p^k> = - |g^k>$$

If $\|g^k\| < 0$, while the Hessian is indefinite, the descent direction is computed from

$$L' \, |p^k\rangle_j = |e^k\rangle \, ,$$

where $|e_j\rangle$ is the j'th column of the unit matrix, and j is the number of the row at which the Hessian was modified in order to make it positive definite. The proof that this $|p^k\rangle$ is in fact a descent direction for the case $\|g^k\| = 0$ and indefinite Hessian is given by Gill, Murray and Picken (1972).

We have written FORTRAN IV routines on basis of the ALGOL procedure MNA (Gill et al. 1972). In our translation of the procedure we have introduced some modifications necessary to conform to the way in which our Hessian matrix is stored. In the original ALGOL version the Hessian is stored row by row in lower triangular from omitting diagonal elements which are stored in a separate array. In our programmes the Hessian is stored row by row in upper triangular form, including diagonal elements. In both versions the Cholesky decomposition is overwritten on the Hessian.

In addition, our implementation of the MNA has the following features:

(1) Storage requirement. Contrary to the common assumption that Newton methods require large amounts of computer storage, our implementation of the MNA requires less storage than the DFP, 4.1 kbyte for MNA and 5.1 kbyte for DFP, not counting the space for common arrays used by both.

(2) Efficiency. The time needed to perform one iteration of the MNA is on average 2-3 times as long as for one DFP iteration. It is compensated, however, by the very fast convergence of the MNA, particularly in the vicinity of the minimum.

(3) Termination criteria. The overall convergence criterion is

$$\|g^k\| < \epsilon 2 = 10(-8).$$

(4) Linear search. The original MNA contains a refined line search based on successive cubic interpolations with safeguards (Gill et al. 1972). In our version we use the line search of the DFP, which is also an efficient cubic interpolation.

Table 5.3 summarises the iteration steps in our three minimisation algorithms.

5.5 The minimisation programme

The three minimisation methods are carried out by a series of sub-routines STEEPD, DAVID, STEPSZ, GAUSS, CHLSKY and LINSOL. These are called from the control programme CONFOR, which is run through once per molecule in the computation. CONFOR performs other functions as well. It reads the lists of packed words prepared by MKLIST, individual molecular control parameters supplied in the input to BRACK, starting coordinates found by BRACK or calculated by REDUCE, all from background files. It also controls the seldomly used calls of TESTER for comparison of numerical and analytical derivatives, and prints lists of the values of all internal coordinates through a call of the subroutine INTOUT, and, if desired, it transforms final coordinates to a specified reference coordinate system by calling the subroutine REFXYZ. In addition, CONFOR performs charge neutral-isation over groups, in a call of subroutine CHARGP, and calculates molecular dipole moments with subroutine DIPOLE. Finally, CONFOR writes a number of arrays, that are to be used in vibrational calculations and in optimisation, on background files. The final coordinates from minimisation are of course also saved, and there are options for producing card image output, which can be used as input to ORTEP and MONSTER.

Table 5.3 Summary of iteration steps in steepest descent, Davidon-Fletcher-Powell and Newton minimisation methods

	Steepest descent	DFP	Modified Newton
1	Initialise. Set $k = 0$. Evaluate $V(\underline{x}^o)$ at the given initial point $|x^o>$		
2	Evaluate the gradient $|g^k>$ at the current point		
3		Compute matrix using recursion formula $H^o = I$	Compute Hessian G^k
4	Set $|s^k> = -|g^k>$	Compute $|s^k> = -H^k|g^k>$	Solve the system of linear eqs. $G^k|s^k> = -|g^k>$
5	Normalise $|s^k>$		Normalise $|s^k>$
6		Compute $\delta^k = \dfrac{<s^k|g^k>}{\|g^k\|}$	Compute $\delta^k = \dfrac{<s^k|g^k>}{\|g^k\|}$

7	If $\|g^k\| < \tau$ terminate, else continue	If $\|g^k\|<\tau 2$ and $\delta^k<\tau 1$ terminate, else	
		If $<s^k\|g^k> > 0$ set $\|s^k> = -\|s^k>$ and $H^k = I$	If $<s^k\|g^k> > 0$ set $\|s^k> = -\|s^k>$

8	Compute the current descent step length, α^k, using one-dimensional search

9	Compute descent step: $\|\Delta x^k> = \alpha^k\|s^k>$

10	Perform descent step: $\|x^{k+1}> = \|x^k> + \|\Delta x^k>$

11	Evaluate $V(\underline{x}^{k+1})$

12	If $V^k - V^{k+1} < \varepsilon 1$ and $\|x^k\| < \varepsilon 2$ terminate, else set $k = k + 1$ and GO TO step 2

5.6 Concluding remarks

5.6.1 Minimisation methods in conformational analysis

The analytical solution of the minimisation problem would have con-
sisted in the solution of a set of non-linear equations:

$$\frac{\partial V}{\partial x_i} = 0 \quad , \; i = 1,2,\ldots,n$$

which is a formidable task due to the complexity of the function
$V(\underline{x})$. This approach would require intolerable simplifications of
$V(\underline{x})$. The only feasible approach to the problem of minimisation of
$V(\underline{x})$ is the use of a numerical method. From our experience with the
three numerical minimisation methods described above we can report
the following general observations.

(1) The best way to minimise the molecular potential energy of a
conformation which represents a very poor guess of the minimum is to
start with the steepest descent method. It is very fast in the
initial iterations and leads rapidly to the region of the minimum,
but it slows down considerably as the minimum is approached asympto-
tically. Therefore only comparatively few iterations should be used,
say, 10-50. Further minimisation should be done with the modified
Newton method.

(2) The steepest descent algorithm is stable and usually yields
conformations that are geometrically more reasonable than those obtained when the DFP is applied to the same very unrealistic conformation.

(3) The availability of analytically computed second order partial derivatives of $V(\underline{x})$ is a necessary condition for the maximum efficiency of a Newton method. The other prerequisite is a good guess of the minimum, such as the result of a steepest descent minimisation. Although the modern Newton methods, including our version of the MNA, are stable also far from the minimum, they require too many iterations and therefore too much time.

(4) Between the two extremes we have the DFP, which has some good characteristics of both the steepest descent (relatively fast initial steps) and the Newton method (quadratic convergence). It is most profitably used in situations neither too far from nor too close to the minimum; if the conformation is too far from minimum, steepest descent is much faster, and if it is too close, Newton is more efficient, because it takes quite a number of steps for DFP to make H a good approximation to the inverse of the Hessian. The use of DFP is indicated when, say a crystal structure analysis is available for input. For small problems, up to fifty dimensions, the DFP may be more efficient than the steepest descent or Newton methods.

5.6.2 Local versus global minimum

As we have already emphasised, the molecular potential energy as a function of cartesian atomic coordinates is an extremely complicated one usually having multiple local minima. Some of these minima may correspond to stable conformations. In conformational analysis, therefore, we are faced with additional problems: to determine if the minimum we have found is unique, and if not, to find other local minima and to determine the global or absolute minimum.

Unfortunately, none of the numerical methods of minimisation can solve the problem of finding the global minimum in a definite way. The most obvious way to find the global minimum is to begin a new search at a number of points in conformational space chosen either at random or over a fixed grid of values of the independent variables and to compare the function values at various local minima (McCormick 1972). In principle, this method reqires an enormous number of function evaluations. In conformational analysis the problem can be reduced to a size tractable in practice by elimination of certain regions of conformational space on the basis of a study of molecular models, which admittedly sometimes may be risky. Another possibility is to perform the search in a selected subspace of fewer dimensions. Recent examples of exhaustive searches for all local minima, and therefore, for the global minimum, are the studies of Scheraga and co-workers who minimised up to 20,000 initial conformations of N-acetyl-N'-methyl amides of animoacids to find a total of four local minima (Lewis et al. 1973), and 2,160 initial conformations of cyclo-hexaglycyl to find a total of nine local minima (Go and Scheraga 1973).

More rigorous approaches are based on numerical solution of non-linear simultaneous equations, by integrating the related system of differential equations (Branin 1972; Branin and Hoo 1972).

In the application of global minimisation techniques to conformational analysis the pioneering work of Scheraga and coworkers (reviewed by Scheraga 1971) is noteworthy.

It may be mentioned here that the related problem of finding transition states has been treated in the CFF context by Ermer (1975).

5.6.3 False minima

Another problem in energy minimisation that has been repeatedly emphasised is that of false minima. In chemical literature false minima are commonly identified with local minima. This is misleading, since multiple local minima are quite common, particularly in large molecules, and can be studied experimentally. For example, a local minimum may have a higher statistical weight than the global one, and may therefore correspond to the actual predominant conformation of a molecule (Scheraga 1971).

False minima are either various stationary points other than true minima or artefacts of inadequate minimisation techniques. Their occurrence can be practically avoided by using efficient minimisation techniques with carefully chosen termination criteria.

A true minimum is identified if the minimisation performed from any of the points obtained by a random search in the vicinity of an achieved minimum converges to the minimum, or if the energy surface around that minimum is well approximated by a quadratic function.

We frequently use a small programme for shaking a molecule if it refuses to go to a proper minimum or if we by some other means have the slightest suspicion that there might be a lower minimum close to one found.

6 VIBRATIONAL CALCULATIONS

Kjeld Rasmussen and Klavs Kildeby

In this chapter we present our implementation of the solution to the
vibrational problem, giving the eigenfrequencies of vibration, and
three types of analysis of normal modes. The exposition is preceded
by a summary of the solution of the vibrational equations of motion, formulated in cartesian coordinates. This method, which is
most rational for large computers, seems to be taking precedence
over the classical Wilson method. It is being used by, among others,
Lifson and Warshel (1968) and Gwinn (1971).

6.1 The vibrational problem

The vibrational equations of motion, in Lagrangian formalism, are

$$\frac{\partial V}{\partial q_i} + \frac{d}{dt}\left(\frac{\partial T}{\partial \dot{q}_i}\right) = 0,$$

where V and T are the potential and the kinetic energy of the
system, t time, and $\dot{q}_i$ the time derivative of the generalised
coordinate q_i.

The entire geometry of a molecule containing N atoms, including its
position and orientation in space, is most rationally given by a
vector of 3N dimensions, whose elements are the cartesian atomic
coordinates.

In terms of these, the kinetic energy may be written as

$$T = \frac{1}{2}\sum_{i=1}^{N}(m_i \dot{x}_i^2 + m_i \dot{y}_i^2 + m_i \dot{z}_i^2),$$

and the potential energy in the neighbourhood of equilibrium as

$$V = V_{eq} + \sum_{i=1}^{3N} \left(\frac{\partial V}{\partial x_i}\right)_{eq} x_i + \frac{1}{2} \sum_{i,j=1}^{3N} \left(\frac{\partial^2 V}{\partial x_i \partial x_j}\right)_{eq} \delta x_i \delta x_j + \ldots$$

The energy zero may be chosen at V_{eq}, which will do away with the first term of the series. At equilibrium, all $\frac{\partial V}{\partial x_i}$ are identically zero, which leaves us with the second and higher order terms. Confining ourselves to small displacements δx_i, we neglect all higher order terms.

We now write the potential energy of vibration as

$$V = \frac{1}{2} \sum_{i,j=1}^{3N} f_{ij} x_i x_j ,$$

where $f_{ij} \equiv \frac{\partial^2 V}{\partial x_i \partial x_j}$, and where x_i from now on will be a short-hand for the cartesian displacement coordinates δx_i. f_{ij} are force constants, though not in the same way as in the usual treatments of vibrational analysis. Rather, each f_{ij} is an individual force constant in a particular molecule, derived for its equilibrium conformation from the force field chosen.

As $\delta \dot{x}_i = \dot{x}_i$, the kinetic energy can be formulated in a similar way as

$$T = \frac{1}{2} \sum_{i=1}^{3N} m_i \dot{x}_i^2 ,$$

where it is to be remembered that the m_i's now come in triplets. In matrix form we have

$$2T = \underline{\dot{x}}'M\underline{\dot{x}}$$

and

$$2V = \underline{x}'F\underline{x}$$

where $\underline{x}$ is the vector of cartesians; $\underline{x}'$ its transpose; M the diagonal matrix of atomic masses, each mass appearing thrice; and F the force constant matrix, which is symmetric as $f_{ij} = f_{ji}$.

In mass-weighted cartesian displacement coordinates

$$\underline{q} = M^{\frac{1}{2}}\underline{x}$$

the energy terms are

$$2T = \underline{\dot{q}}'\underline{\dot{q}}$$

and

$$2V = \underline{q}'M^{-\frac{1}{2}}FM^{-\frac{1}{2}}\underline{q} = \underline{q}'H\underline{q}.$$

Insertion into the equations of motion yields the set of equations

$$\underline{\ddot{q}} + H\underline{q} = \underline{0}.$$

The solutions, of the form

$$q_{ik} = l_{ik} \cos(\omega_k t + \alpha)$$

$$\ddot{q}_{ik} = -\omega_k^2 q_{ik} = -\lambda_k q_{ik}$$

or

$$\underline{q}_k = \underline{l}_k \cos(\omega_k t + \alpha)$$

give, on insertion,

$$(H - \lambda_k E)\underline{l}_k = \underline{0}$$

or, collecting all 3N equations,

$$HL = L\Lambda.$$

Λ is the diagonal matrix of eigenvalues of H, and L, whose columns are $\underline{l}_k$, is the matrix of eigenvectors $\underline{l}_k$. L is unitary, and is found by diagonalisation of H:

$$H=L\Lambda L'.$$

Note that, in contrast to Wilson's method, no G matrix enters, and only one diagonalisation is required.

The eigenvalues are easily extracted in any suitable unit of energy or frequency.

6.2 Normal coordinates

We now have the normal coordinates Q_i, which diagonalise the potential energy of vibration:

$$2V = \underline{q}'L\Lambda L'\underline{q} = \underline{Q}' \Lambda \underline{Q}$$
$$2T = \underline{\dot{Q}}'\underline{\dot{Q}}$$
$$\underline{Q} = L'\underline{q} = L'M^{\frac{1}{2}}\underline{x}$$
$$\underline{x} = M^{-\frac{1}{2}}L\underline{Q}$$

The normal coordinates, which characterise the motions of all atoms in a molecule, for each eigenvalue of vibration, can thus be written as linear combinations of cartesian displacement coordinates.

$\underline{l}_k$ should be interpreted as the amplitudes of $\underline{q}_k$, the mass-weighted cartesian displacements forming up the k'th normal coordinate $\underline{Q}_k$. The cartesian displacements are thus found as $M^{-\frac{1}{2}}\underline{l}_k$. These may be used to visualise the atomic movements in any normal mode, for instance through a plotter programme.

Further transformation into internal coordinates will facilitate assignments, both with respect to symmetry types, and into classes of group, zone and delocalized frequencies (Jones et al. 1972).

Transformation from cartesian displacement to internal displacement coordinates is done using partial derivatives of internals with respect to cartesians (reverting to δ for displacement):

$$\delta r_i = \sum_{j=1}^{3N} \frac{\partial r_i}{\partial x_j} \, \delta x_j$$

or

$$\underline{\delta r} = B \underline{\delta x} \, ,$$

$$B_{ij} = \frac{\partial r_i}{\partial x_j} \, .$$

B is a rectangular matrix having 3N columns and a number of rows equal to the number of internals.

Amplitudes of internal displacements are thus found as $BM^{-\frac{1}{2}}\underline{l}_k$, or as the elements of $BM^{-\frac{1}{2}}L$.

6.3 Programme VIBRAT

The subroutine VIBRAT, which is run through once per molecule, reads the DD-matrix (the F-matrix) from a background file, into which CONFOB has written it. It then forms the mass-weighted H-matrix and diagonalises it.

The diagonalisation is done by a call of subroutine EIGEN, which uses first a Householder tridiagonalisation and then a QR-algorithm (Gourlay and Watson 1973). This programme is much lengthier than the classical Jacobi rotation, but the smaller number of rotations makes it much faster to run. Our tests (Gregory and Karney 1969, Wilkinson 1965) indicated that it pays, in terms of computer time, to use the Householder + QR method for problems of more than about 30 independent variables. Further our Householder programme reproduces exactly Wilkinson's eigenvectors for degenerate eigenvalues, which the Jacobi programme does not. The subroutine EIGEN is a modification of a version developed at the Weizmann Institute Computer Center. Comparison was done with a standard Jacobi subroutine of the IBM Scientific Subroutine Package.

After diagonalisation, VIBRAT calculates frequencies from eigenvalues, and, if specified, normal coordinates in terms of cartesian or internal displacement coordinates.

The B matrix is taken from a background file, where it has formerly been written by CONFOR. The elements are calculated and organised by the subroutines xxxP, xxx = BOND, THETA, PHI, NBOND and UREY.

A subroutine SYMANA sorts the internal displacement coordinates for each normal coordinate and prints them in a way which allows the reader to determine, on inspection, its symmetry type and its character in respect to group, zone or delocalised vibration. The subroutine was written by Oliver Jacobsen during an undergraduate course of molecular spectroscopy.

A subroutine INTENS calculates rough infrared intensities in a simple classical way, treating each normal coordinate as a vibrating dipole. This will work, of course, only when fractional charges are used in the potential energy function.

6.4 Practical considerations

The use and mode of function of VIBRAT is governed by an individual control parameter specified in the input to BRACK, which means that it applies to a specific molecule. It will be understood from Section 6.1 that it is meaningless to calculate frequencies before the molecule has attained a conformation of minimum energy.

A practical check on the calculations is given by the six lowest eigenvalues, which should be zero for a nonlinear molecule. Also the eigenvectors provide valuable checks. Of the six external degrees of freedom, three should be translations, and three rotations. The translations are easy to identify in the list of cartesian displacement coordinates. Lastly, all elements of the six last internal displacement eigenvectors, corresponding to the external degrees of freedom, should be zero.

7 OPTIMISATION OF ENERGY PARAMETERS

Steen Melberg and Kjeld Rasmussen

In this chapter we shall be concerned once again with the very basis
of the CFF concept. It will be recalled that this concept may be
formulated as: choice of analytical forms of potential energy
functions for a family of molecules; selection of initial values of
the parameters of the energy functions; calculation of any desired
observable property of each molecule; optimisation of the energy
parameters to provide a better fit between calculated and measured
observables. We shall treat here this concluding step.

The algebraic derivation is modified from that of Lifson and Warshel
(1968, 1970), who chose a least squares algorithm with weighting of
observables, and with constraints on parameter variation introduced
through a Lagrange multiplier.

We still use linearised least squares with weighting of observables,
but in solving the normal equation we use QR decomposition and
Givens transformations which give a numerically stable algoritm.

7.1 The basic algorithm

Let $\underline{x}$ be a vector whose elements are the current values of those
energy parameters we want to optimise:

$$\underline{x} = |x_1, x_2, \ldots, x_m, \ldots, x_{nopt}>,$$

where nopt is the number of parameters to be optimised.

$\underline{\delta x}$ will be a vector whose only non-zero element is δx_m, a slight change in the value of x_m:

$$\underline{\delta x}_m = | 0, 0,\ldots, \delta x_m ,\ldots, 0\rangle$$

Accordingly,

$$\underline{\delta x} = \sum_{m=1}^{nopt} \underline{\delta x}_m .$$

Let y be a vector whose elements are those values of the observables we want to optimise on:

$$\underline{y}^{meas} = | y_1, y_2 ,\ldots, y_k ,\ldots y_{ntot}^{meas} \rangle$$

$$\underline{y}^{calc} = | y_1, y_2 ,\ldots, y_k ,\ldots y_{ntot}^{calc} \rangle$$

where ntot is total number of observables to be optimised, counting over all molecules in the set.

The differences between measured and calculated values will be

$$\Delta y_k = y_k^{meas} - y_k^{calc} .$$

Accordingly,

$$\underline{\Delta y} = | \Delta y_1, \Delta y_2 ,\ldots, \Delta y_k ,\ldots, \Delta y_{ntot} \rangle.$$

The problem is now, given $\underline{x}$, to find a $\underline{\delta x}$, which will make $\underline{\Delta y}$, or rather its weighted norm, smallest possible.

In the present exposition, y_k may be any bond length, valence angle, torsional angle or internal frequency of any molecule in the set being treated. In earlier work (Lifson and Warshel 1968, 1970), y_k could be a thermodynamic function or a unit cell dimension. In future work, we intend to include these, as well as rotational constants, infrared intensities, and proton-proton spin-coupling parameters.

The above statement serves to illustrate that $\underline{\Delta y}$ in general will be an extremely complex function of $\underline{x}$. We shall now linearise it. Let us expand $\underline{\Delta y}$ in a Taylor series around the point $\underline{x}$:

$$\underline{\Delta y}(\underline{x} + \underline{\delta x}) = \underline{\Delta y}(\underline{x}) - Z\underline{\delta x} + \ldots,$$

where

$$Z_{km} = -\frac{\partial \Delta y_k}{\partial x_m} = -\frac{\partial (y_k^{meas} - y_k^{calc})}{\partial x_m} = \frac{\partial y_k^{calc}}{\partial x_m},$$

$$k = 1, \ldots, ntot; \quad m = 1, \ldots, nopt$$

Assuming well-behaved functional dependence of $\underline{\Delta y}$ on $\underline{x}$, and confining ourselves to small $\underline{\delta x}$, we shall neglect higher-order terms.

y_k^{meas} will have very different experimental uncertainties. They are therefore weighted by the inverse of their absolute uncertainties. These reciprocals will be called P_k; they form a diagonal matrix P of order ntot. A more logical weighting scheme was recently proposed by Wiffen (1976); this will be tried in our programme.

We now have

$$\underline{r} = \underline{\Delta y}(\underline{x} + \underline{\delta x}) = P\underline{\Delta y}(\underline{x}) - Z\underline{\delta x}$$

The squared residual r^2 is given by

$$
\begin{aligned}
r^2 &= \|\underline{r}\|^2 = (P\underline{\Delta y}(\underline{x}) - Z\underline{\delta x})'(P\underline{\Delta y}(\underline{x}) - Z\underline{\delta x}) \\
&= (\underline{\Delta y}(\underline{x})'P' - \underline{\delta x}'Z')(P\underline{\Delta y}(\underline{x}) - Z\underline{\delta x}) \\
&= \underline{\Delta y}(\underline{x})'P'P\underline{\Delta y}(\underline{x}) - \underline{\Delta y}(\underline{x})'P'Z\underline{\delta x} - \underline{\delta x}'Z'P\underline{\Delta y}(\underline{x}) + \underline{\delta x}'Z'Z\underline{\delta x}
\end{aligned}
$$

We therefore seek $\underline{\delta x}$ as the solution to the set of nopt equations

$$\frac{\partial}{\partial \delta x_m}(\|\underline{r}\|^2) = 0$$

By differentiating r^2 through $\underline{\delta x}'$ we get

$$- Z'P\underline{\Delta y}(\underline{x}) + Z'Z\underline{\delta x} = 0$$

or

$$Z'Z\underline{\delta x} = Z'P\underline{\Delta y}(\underline{x}).$$

This derivation differs both from the original one of Lifson and Warshel (1968) and from that of Ermer and Lifson (Ermer 1976).

Solving these equations numerically has serious drawbacks, because the crossproduct matrix Z'Z is often ill-conditioned.

If the decomposition

$$Z = QR$$

is available, where Q is an ntot * nopt matrix, whose columns are orthonormal, and R is an nopt * nopt upper triangular matrix, then

$$R'Q'QR\delta x = R'Q'P\Delta y(x).$$

As

$$Q'Q = E,$$
$$R'R\delta x = R'Q'P\Delta y(x).$$

Since R is non-singular if Z'Z is,

$$R\delta x = Q'P\Delta y(x)$$

This triangular system can therefore be solved for δx without forming the crossproduct matrix.

The QR decomposition can be obtained by applying a sequence of Givens transformations (plane rotations) to Z and $\Delta y(x)$ (Wilkinson 1965).

Using the algoritm proposed by Gentleman (1973) δx can be obtained without calculating square roots, which are normally needed for Givens transformations.

As δx is found, the expected new differences between measured and calculated values can be determined from

$$\Delta y(x + \delta x) = \Delta y(x) - Z\delta x$$

or

$$\Delta y(x^{new}) = \Delta y(x^{old}) - Z\delta x^{old}.$$

The next iteration will be of the form

$$\underline{\Delta y}(\underline{x}^{\text{new}} + \underline{\delta x}) = \underline{\Delta y}(\underline{x}^{\text{new}}) - z\underline{\delta x}$$

and so on, until the determined vector difference $\underline{\Delta y}(\underline{x} + \underline{\delta x})$ is sufficiently small.

7.2 The partial derivatives

The elements of the Z matrix, which are the partial derivatives of observables with respect to energy function parameters, are very difficult to obtain, as the internal coordinates $\underline{r}$ are extremely complicated functions of the energy parameters $\underline{x}$. Derivatives of internal frequencies are not nearly as complicated. One difficulty, common to both classes, is that the Z elements should really be calculated at a new equilibrium after each differential change in a parameter. If the derivatives were to be calculated purely numerically, this would entail a full minimisation cycle per molecule per parameter to be optimised, which would be forbidding in terms of computer time. This diffucully must therefore be circumvented by an approximation.

7.2.1 Internal coordinates

In this section $\underline{y}$ will denote internal coordinates, $\underline{c}$ cartesian coordinates and $\underline{x}$ energy parameters.

The elements of Z are

$$Z_{km} = \frac{\partial y_k}{\partial x_m} = \sum_{l} \frac{\partial y_k}{\partial c_l} \frac{\partial c_l}{\partial x_m}$$

or

$$z_m = B \frac{\partial c}{\partial x}$$

where the B matrix is the same as that used in Section 6.2.

We now have to find the derivatives

$$\partial c/\partial x_m = |\partial c_1/\partial x_m, \partial c_2/\partial x_m, \ldots, \partial c_e/\partial x_m, \ldots >.$$

They are defined as

$$\frac{\partial c_o(x)}{\partial x_m} = \lim_{\delta x_m \to 0} \frac{c_o(x+\delta x) - c_o(x)}{\delta x_m}$$

where the subscript o denotes equilibrium conformation.

In principle, c_o is a known function of x and therefore also of δx_m, through the equilibrium conditions

$$\nabla V(c_o(x); x) = 0$$

$$\nabla V(c_o(x+\delta x_m); x + \delta x_m) = 0.$$

In Chapter 5, $c_o(x)$ was found from the Taylor expansion

$$\nabla V(c(x); x) = \nabla V(c_o(x); x) + F(c_o(x); x) \delta c$$

and the above condition

as

$$\underline{c}_o(\underline{x}) = \underline{c}(\underline{x}) - F^{-1}(\underline{c}_o(\underline{x}); \underline{x}) \nabla V(\underline{c}(\underline{x}); \underline{x})$$

where $\underline{c}$ is an arbitrary initial conformation.

$F(\underline{c}_o(\underline{x}); \underline{x})$ was approximated with $F(\underline{c}(\underline{x}); \underline{x})$, and the equation was solved by iteration.

Analogously, $\underline{c}_o(\underline{x}+\delta\underline{x}_m)$ may be found from

$$\nabla V(\underline{c}(\underline{x}+\delta\underline{x}_m); \delta\underline{x}_m) = \nabla V(\underline{c}_o(\underline{x}+\delta\underline{x}_m); \delta\underline{x}_m) + F(\underline{c}_o(\underline{x}+\delta\underline{x}_m); \delta\underline{x}_m)\delta\underline{c}$$

and the equilibrium condition as

$$\underline{c}_o(\underline{x}+\delta\underline{x}_m) = \underline{c}(\underline{x}+\delta\underline{x}_m) - F^{-1}(\underline{c}_o(\underline{x}+\delta\underline{x}_m); \delta\underline{x}_m) \nabla V(\underline{c}(\underline{x}+\delta\underline{x}_m); \delta\underline{x}_m).$$

As the arbitrary initial conformation $\underline{c}$ we may choose the conformation we know from minimisation, $\underline{c}_o(\underline{x})$:

$$\underline{c}_o(\underline{x}+\delta\underline{x}_m) = \underline{c}_o(\underline{x}) - F^{-1}(\underline{c}_o(\underline{x}+\delta\underline{x}_m); \delta\underline{x}_m) \nabla V(\underline{c}_o(\underline{x}); \delta\underline{x}_m).$$

Approximating F computed at equilibrium with F at the initial conformation, just as before, we get

$$\underline{c}_o(\underline{x}+\delta\underline{x}_m) = \underline{c}_o(\underline{x}) - F^{-1}(\underline{c}_o(\underline{x}); \delta\underline{x}_m) \nabla V(\underline{c}_o(\underline{x}); \delta\underline{x}_m)$$

and, as a difference quotient,

$$\frac{\partial c(x)}{\partial x_m} = \frac{c_o(x+\delta x_m) - c_o(x)}{\partial x_m}$$

$$= - F^{-1}(c_o(x); \delta x_m) \nabla V(c_o(x); \delta x_m)/\delta x_m .$$

This means that both the gradient and the Hessian matrix are calculated at the equilibrium conformation as found with unchanged
energy parameters, but now with one parameter changed at a time. The
set of linear equations can be solved by standard methods.

Lifson and Warshel (1968) used another approximation for F,
$F(c_o(x))$. Therefore they computed the Hessian matrix only once per
molecule per iteration step in the optimisation, whereas we do it
nopt times per molecule per iteration. However, the programme MOLEC
and all its subprogrammes must anyway be called because the gradient
is needed, and our algorithm avoids many transports of the large
Hessian from background memory. In addition, we believe that our
method is more accurate.

7.2.2 Internal frequencies

In this section, $\underline{y}$ denotes internal frequencies.

The frequency subset of the Z matrix is found through the eigenvalues of vibration

$$\lambda_k = (const * y_k)^2$$

as

$$\frac{\partial y_k}{\partial x_m} = \frac{\partial y_k}{\partial \lambda_k} \frac{\partial \lambda_k}{\partial x_m} = \frac{1}{2(\text{const}) y_k^2} \frac{\partial \lambda_k}{\partial x_m} .$$

$\dfrac{\partial \lambda_k}{\partial x_m}$ are found from the secular equation as

$$\partial \lambda_k / \partial x_m = \delta q'_{-k} \; \partial /\partial x_m \; \delta q_{-k} = \delta q'_k \; U' \; M^{-\frac{1}{2}} \; \partial F / \partial x_m \; M^{-\frac{1}{2}} \; U \; \delta q_{-k} .$$

The derivative of the Hessian is, by definition,

$$\frac{\partial F}{\partial x_m} = \lim_{x_m \to 0} \frac{F(C_{-o} (x+\delta x_m); \delta x_m) - F(C_{-o} (x); x)}{\delta x_m}$$

The first term in the denominator is known from the previous section, and the second term is the equilibrium Hessian.

7.3 Implementation of the optimisation

The entire set of optimisation programmes is organised as Section VII of our system.

7.3.1 Reading of experimental data

The subroutine RDEXP is called once per job by the MAIN programme. It is run through once per molecule, and for each molecule once per type of observable, presently conformations and vibrations. The following data are read: the list number of each internal coordinate or frequency to optimise on; its experimental value; its experimental uncertainty. The routine counts the number of data, and stores them on a background file.

7.3.2 Organisation of the optimisation

The programme OPTIM controls the optimisation. Through calls of subroutine NPAR it changes by small amounts the values of those parameters that are to be optimised, one by one and through subroutine BUILDZ the elements of the Z matrix are built up, one row per parameter to be optimised. The rows are written on a background file.

The Z matrix is put together by the subroutine ZMATRX; and the $p\Delta y$ vector is constructed by BUILDY. The least sqares algorithm is performed by LSTSQR. OPTIM prints the new parameter values, and updates them in the system by a call of NPAR. OPTIM also calculates and prints various statistical measures.

Control is then returned to the MAIN programme and the whole series of conformational and vibrational calculations and optimisation may be repeated.

8 DEVELOPING A FORCE FIELD

Kjeld Rasmussen

The relevance of the development and use of a programming system as large and complicated as the one described in previous chapters will inevitably be questioned. This is not the place for a philosophical discussion of the virtue or the waste in doing conformational analysis. If the soundness of such research is accepted, however, a crucial question still remains: What is to be gained from investing so much effort in programme development?

The answer, in the author's opinion, lies in a parallelism to the work of the early atomic spectroscopists. The reduction of series of atomic spectral lines to systems of terms was, just as much as the then new quantum theory of Planck and Einstein, a necessary prerequisite to Bohr's postulates of 1913, which started the development of theoretical chemistry and physics as we know them today.

Those who nowadays strive to develop potential energy functions are certainly not gifted with the imagination and intuition of Rydberg and Ritz. Nonetheless, we should set ourselves an analogous goal: To parametrise a vast amount of experimental information using as simple expressions and as few parameters as consistent with a reasonably accurate description. It would be wrong to imply that we will eventually provide a solid basis for a new revolutionary theory; but we may at least hope to be able to inspire experimentalists to test our predictions, based as they will be on analysis of empirical evidence; and to challenge theorists to provide us with better insight in the structure and interactions of matter.

Given a programme and a manual, then, how does one set out to develop a set of energy functions, with sensible parameters, for the description of structures, vibrations and thermodynamics of a particular class of compounds? A clear-cut answer cannot be given, but in the following sections a discussion of principles and a few examples will be given. They reflect the author's points of view and are by no means as objective as are, hopefully, the earlier parts of this exposition.

Those seeking comments and bibliographies to the entire field are referred to the excellent recent reviews of Altona and Faber (1974), Allinger (1976), and Ermer (1976).

8.1 The concept of energy functions

From the discussion in Chapter 4 it will be clear that the splitting of the potential energy of atomic motion into bond, angle and non-bonded terms etc., is a gross simplification. The only justification of this approach is that it works, so let us see if we can understand why.

The classification of interactions as bonded and non-bonded reflects, of course, the traditional chemical ball and spoke visualisation of the structure of matter. It reflects also the quantum chemical description, bonds corresponding to valence electron clouds shared by two or more atomic cores. We shall discuss in some detail these two classes of interaction.

8.1.1 Bonded interactions

8.1.1.1 Bonds

A good bond energy function should represent both the steep repulsion between two atomic cores when they are brought too close together, and the not nearly so steep attraction between the electron cloud of the bonding region and the two cores when they are pulled apart, falling off to zero at large separation.

These criteria are met by the Morse function. In spite of this it is not used extensively in conformational analysis, for two simple reasons. One is merely that parameters are available for only very few bonds. The other is that a parabola is a fair approximation in a small interval about the minimum. If bonds are not too strained from, say, overcrowding, it is, therefore, sensible to use a simple parabola, which has the advantage of requiring only two parameters.

Using a parabola as bond function does not mean that the actual force is harmonic. Take a methane molecule and compress one bond. The energy increases more than what is due to the harmonic function, because of the non-bonded repulsion between the singular hydrogen atom and the other three. Conversely, when the bond is stretched, the energy does not increase parabolically, because of London attraction between the hydrogen atoms. If Coulomb terms are added to the non-bonded interactions, this tendency of anharmonising the bond is strengthened further.

We thus see that, in polyatomic molecules, bond stretching becomes anharmonic even if a harmonic bond energy function is used. How good the representation is depends crucially on the treatment of non-bonded interactions, to which we shall return below.

As mentioned above, a harmonic bond function requires two parameters and a Morse function three. It should be possible to reproduce an anharmonic function like the Morse with two or at most three inverse power terms. Such a function would require two or three parameters and be computationally very simple, and would merit a closer study.

8.1.1.2 Torsions

In Chapter 4 the concept of bond torsion was treated at some length. Here we shall just stress that if torsion is considered a bond property, it vanishes for a pure sigma bond, and for a double bond a parabola would probably be more realistic than the traditional trigonomic functions. If on the other hand torsion is considered a non-bonding property, as between opposite C - H bonding regions in ethane or ethene, then the stiffness of a double bond is not adequately treated. It is suggested that the first concept be used, leaving the rotational barrier around single bonds to be treated entirely with non-bonded terms.

8.1.2 Non-bonded interactions

8.1.2.1 Atom-atom interactions

Also this aspect was treated at length in Chapter 4; in particular, a discussion of the more common functions was given. Here we shall reflect on which interactions in a molecule should actually be included.

One would be tempted to include interactions between just such atoms as are able to 'see' each other, without their 'line of sight' being blocked by other atoms or bonding regions. Logical though such an approach would be, it presents formidable problems in programming, and, to the author's knowledge, has never been tried.

Another problem in enumerating the interactions is whether or not to include interactions between two atoms bound to the same third atom. They are usually not included, or at any rate not treated in the same way as other non-bonded interactions, although the characteristic distances are not shorter, or are even longer, than many 'ordinary' non-bonded distances in complicated molecules. Such common treatment of all non-bonded interactions should definitely be tried.

Until now, interactions between two atoms bound to a third have either been left out, or have been treated with a Urey-Bradley term. There is really no logical reason for this special treatment.

Coulomb interactions between fractional atomic charges in the monopole approximation are used by many groups. These very far-reaching interactions are definitely important, though a lot more work remains to be done. It is for instance quite conceivable that some other function than the inverse first power should be used, as a way of approximating the change of dielectric constant from the vacuum value at close distance to some much higher value for two charges on opposing corners of a molecule. It is not even obvious that the concept of a bulk dielectric constant applies in a molecule. The whole subject of electrostatic interactions in molecules calls for theoretical investigation.

8.1.2.2 Geminal interactions

Geminal or 1,3-interactions operate between the charge clouds of two bonding regions protruding from an atom. Here is one particular point where we should hope for the theorists to provide us with some better analytical description of interaction than the traditional harmonic angle term.

8.2 Examples

Personal experience is of course the best guide to the development of potential energy functions and to the choice of parameters. For the beginner, but certainly also for the experienced researcher, much help may be had from previous work. We shall here attempt to give some guidance by means of a series of examples. All units are chosen to give the energy in kcal/mol, with lengths in Å and angles in rad.

8.2.1 Pre-CFF, cycloalkanes

Bixon and Lifson (1967) got surprisingly good results using a rather primitive set of energy functions, with judiciously selected parameters.

Their choices are fully explained in the paper. Note that the parameters with dimensions of force constants are twice the values given; for instance $1/2 \ K(C-C) = 300$.

```
C-C        bonds:     E = 300(b-1.533)²

C-H        bonds:     none


C-C-C      angles:    E = 80(θ-1.96)²

C-C-H      angles:    none

H-C-H      angles:    none


C-C-C-C torsions:     E = 1.7(1+cos3φ)

other    torsions:    none


H---H    non-bonds:   E = 10(4)exp(-4.60r)-49.2/r⁶

other    non-bonds:   none
```

8.2.2 Original CFF, n- and cycloalkanes

Lifson and Warshel (1968) made a penetrating study of many energy functions, which should be read by anyone interested in conformational calculations.

Each set of functions was optimised. We cite in Table 8.1 their final values for a rather complicated modified Urey-Bradley force field:

$$E = \sum 1/2\ K(b-b0)^2 + \sum [1/2\ K(\theta-\theta 0)^2 + K'(\theta-\theta 0)]$$

$$+ \sum 1/2\ K(1+\cos 3\phi) + \sum [1/2\ F(r-r0)^2 + F'(r-r0)]$$

$$+ \sum [\epsilon(r*/r)^{12} - 2\epsilon(r*/r)^6 + ee/r]$$

Table 8.1 CFF of Lifson and Warshel (1968)

type	1/2 K	b0,θ0	K'	1/2 F	F'	r0
C-H	286.9	1.099				
C'-H	314.4	1.099				
C-C	111.0	1.455				
C-C-C	22.0	1.911	-7.48	37.31	-1.547	2.5
C-C-H	26.79	1.911		43.61	-0.746	2.2
C-C'-H	24.3	1.911				
H-C-H	38.14	1.911		2.900	-0.104	1.8
C-C-C-C	1.418					

	ε	r*
C---C	0.0196	4.228
C---H	0.0045	2.936

$C' = C$ in methyl, $\varepsilon CH = (\varepsilon CC \varepsilon HH)^{\frac{1}{2}}$, $r*CH = (r*CC + r*HH)/2$,

$$e = \pm 0.144 \text{ el. ch.}$$

8.2.3 CFF, alkane crystals

A variant of the original CFF was optimised on additional data, including crystal properties (Warshel and Lifson 1970). A Lennard--Jones 9-6 function was used instead of the 12-6 function and an angle-torsion interaction term $K'(\theta-\theta0)(\theta'-\theta0')\cos\phi$ was added, and the linear term of the Urey-Bradley function was dropped. The results are cited in Table 8.2.

Table 8.2 CFF of Warshel and Lifson (1970)

type	1/2 K	b0,θ0	K'	1/2 F	r0
C-H	286.4	1.099			
C'-H	310.6	1.102			
C-C	110.3	1.490			
C-C'	110.3	1.467			
C'-C'	110.3	1.444			
C-C-C	15.5	1.911	-6.2	55.0	2.5
C-C-H	25.3	1.911		42.9	2.2
C-C'-H	18.3	1.911		51.7	2.2
H-C-H	39.5	1.911		1.7	1.8
C-C-C-C	1.161		-2.3		
H-C-C-C	1.161		-6.9		
H-C-C-H	1.161		-9.5		

	$\varepsilon^{1/2}$	1/2 r*
C---C	0.4297	1.808
H---H	0.0508	1.774

C' = C in methyl, e = ± 0.11 el. ch.

8.2.4 CPF, amides and lactams

Warshel, Levitt and Lifson (1970) optimised a parameter set on conformations and vibrational spectra of small amides and lactam rings. The original CPF was modified by dropping the linear angle term, and by adding terms $1/2K(1-cos2\phi)$ for torsion around the peptide bond and $1/2K\chi^2$ for out of plane angles at both ends of the peptide bond. The linear Urey-Bradley parameter was locked with $F' = -0.1Fr0$. Table 8.3 cites the results as given in the paper. ε and r^* for atoms other than N and O are not given, neither are fractional charges shown. Also parameters for C-H etc. are notably absent. Schellman and Lifson (1973) optimised a force field for pyrrolidones. It looks very much like a merging of Tables 8.2 and 8.3. Fractional charges are there given as: H on N 0.27, H on C 0.14, O -0.42, N -0.30, K 0.45, C in CH3 -0.42, C in CH2 -0.28, C in CH -0.14.

Table 8.3 CFF of Warshel, Levitt and Lifson (1970)

type	1/2 K	b0,θ0	1/2 F	r0
N-H	405	0.980		
N-K	403	1.278		
K-H	259	1.040		
K-O	595	1.200		
K-C	187	1.470		
C-N	261	1.457		
K-N-H	26.6	2.094	27.9	2.000
K-N-C	54.5	2.094	16.2	2.400
C-N-H	31.4	2.094	26.0	1.791
N-K-O	48.5	2.094	90.0	2.186
N-K-C	33.1	2.094	50.5	2.229
N-K-H	17.5	2.094	43.2	2.100
O-K-C	40.9	2.094	52.0	2.400
O-K-H	22.8	2.094	66.2	2.000
H-C-N	30.1	1.911	41.0	1.900
H-C-K	26.8	1.911	38.4	1.975
C-K-N-C	1.655			
C-K-N-H	1.655			
O-K-N-C	4.487			
O-K-N-H	4.045			
H-C'-K-X	0.500			
H-C'-N-Y	1.500			
CKO, KN	4.04			
HNC, NK	0.69			

	$\sqrt{\varepsilon}$	$1/2\ r*$
N	0.44	1.8
O	0.48	1.5

X = O, N; Y = H, C; K = C of amide bond; C' = C in methyl

8.2.5 CFF, amides

Hagler, Huler and Lifson (1974) optimised Lennard-Jones 9-6 and 12-6 functions plus fractional charges on crystal structure, heat of sublimation and dipole moments of a number of amides. Molecules were treated as rigid bodies. The most important result is that the hydrogen bond can be described by ordinary non-bonded plus Coulomb interactions and thus requires no special treatment, with for instance Stockmayer, Morse or Lippincott and Schroeder functions. Final parameters for the functions

$$AA'/r^{12} - BB'/r^6 + ee'/r \text{ and}$$

$$AA'/r^9 - BB'/r^6 + ee'/r \text{ and in the equivalent forms,}$$

$$\varepsilon[\,(r*/r)^{12} - 2(r*/r)^6\,] + ee'/r \text{ and}$$

$$\varepsilon[\,2(r*/r)^9 - 3(r*/r)^6\,] + ee'/r$$

are cited in Table 8.4

Table 8.4 Non-bonded parameters for amides

atom	12-6 A*10(-3)	B	e	9-6 A*10(-3)	B	e
O	275	502	-0.38	45.8	1410	-0.46
N (NH)	2271	1230	-0.28	86.9	2020	-0.26
N (NH2)	2271	1230	-0.83	86.9	2020	-0.82
K	3022	1340	0.38	12.5	355	0.46
C	1811	532		38.9	1230	
H (C)	7.15	32.9	0.10	0.445	15.0	0.11
H (NH)	0	0	0.28	0	0	0.26
H (NH2)	0	0	0.41	0	0	0.41

	ε	r*	ε	r*
O	0.228	3.21	0.198	3.65
N	0.167	3.93	0.161	4.01
K	0.148	4.06	0.042	3.75
C	0.039	4.35	0.184	3.62
H (C)	0.038	2.75	0.0025	3.54
H (N)	0	0	0	0

8.2.6 Flexible amino acids

Gelin and Karplus (1975) made a study of potential surfaces for
acetylcholine and methylacetylcholine. They used a version of the
CFF programme, but did not optimise the energy parameters. Rather,
they employed, as most user will probably do, a parameter set developed by optimisation and modified it after having examined the
results of trial calculations of conformations.

The energy functions are given in their eq. 1 and the parameter set in their Table 1. Fractional charges were obtained by INDO calculations; unfortunately they are not reported in the paper.

8.2.7 Coordination compounds

Niketić et al. (1973, 1974, 1976) have selected a set of energy functions with parameters for conformational studies on tris-(di-amine) chelate coordination complexes. They adopted harmonic functions for bond stretching and angle bending; Pitzer-type potential with threefold periodicity for rotations around single bonds; and a Buckingham-type function for non-bonded interactions.

They used the torsional function concept of bond torsional energies (see Section 4.3.2) whereby the torsional contribution for a bond is obtained as a sum of bond interactions (nine for an sp3-sp3 bond, etc.) each with an appropriate fraction of the energy parameter value chosen to reproduce the corresponding rotational barrier.

As this would be impractical in vibrational analysis to follow since it entails too many internals, they tried one bond interaction for each torsional angle (the concept of group torsional energies) and found that this modification caused insignificant differences in computed structures and energies.

Bond stretching and angle bending parameters for the hydrocarbon part of the force field were taken from Wiberg's force field (Wiberg 1965; Harris 1966; Gleicher and Schleyer 1967). It was supplemented with harmonic deformations of bonds and angles involving metal and coordinated nitrogen atoms, which were taken from the normal coordinate analyses of ammine complexes of cobalt(III) by Nakagawa and Shimanouchi (1966).

Several sets of non-bonding parameters including those for heteroatoms are currently in use. The choice was the set of parameters for
Buckingham functions developed by Liquori (1969) on the basis of the
second virial coefficients of gases (De Coen et al., 1967), and
tested on a variety of molecules (Liquori et al. 1968).

Torsional parameters (unique value for both C-C and C-N bonds) was
adjusted so that the force field could reproduce the rotational
barrier of about 3 kcal/mol in ethane using the abovementioned
non-bonded functions.

The application of this force field to octahedral tris-bidentate
metal chelate complexes implied some special conditions: (1) Since
geminal interactions were accounted for in the angle bending terms
and therefore not treated explicitly, all non-bonded N...N interactions were omitted. (2) Valence angles defined by ligating atoms
in trans position as well as those between the ligators from different chelate rings were not treated. In this way only three
chelate angles were considered at the octahedral metal atom. (3)
Exclusion of the so called core field potential (non-bonded
interactions involving the central metal atom; in this force field
M...C and M...H contributions) has practically no significance on
the results of force field calculations (see also Dwyer and Searle
1972). This was demonstrated (Niketić et al., 1976) by test
computations in which core field terms were included with the
appropriate parameters for M set equal to those of C.

The parameters used with the energy function

$$E = \sum 1/2\ K(b-b0)^2 + \sum 1/2\ K(\theta-\theta0)^2 + \sum 1/2\ K(1+\cos n\phi)$$

$$+ \sum [\text{Aexp}(-Br) - C/r^6]$$

are shown in Table 8.5.

Table 8.5 Parameters for coordination compounds

type	K	b0,θ0,n
M-N	251.65	2.00
N-C	862.80	1.47
C-C	719.00	1.54
C-H	719.00	1.093
N-H	805.28	1.011
N-M-N	97.784	1.571
M-N-H	28.760	1.911
M-N-C	57.520	1.911
N-C-C	143.80	1.911
N-C-H	93.470	1.911
H-N-H	76.214	1.911
C-N-H	93.470	1.911
H-C-H	74.776	1.911
H-C-C	93.470	1.911
C-C-C	143.80	1.911
X-C-C-Y	2.8	3
X-C-N-Y	2.8	3
X-M-N-X	0.0	12

	A*10 (-4)	B	C
H---H	0.66	4.08	49.2
H---C	3.14	4.20	121.1
H---N	2.81	4.32	99.2
C---N	21.21	4.44	244.0
C---C	23.70	4.32	297.8
N---N	18.64	4.55	200.0
N---H	3.14	4.20	121.1
N---C	23.70	4.32	297.8

X,Y = N, N, C, H

8.2.8 Saccharides

Kildeby, Melberg and Rasmussen (1977) discussed at some length their
selection of energy functions and parameters by modification of sets
developed for other purposes. Their final choice, which gave a fair
description of glucose conformations, is reproduced in Table 8.6.
The parameters correspond to the rather simple energy function

$$E = \sum 1/2\ K(b-b_0)^2 + \sum 1/2\ K(\theta-\theta_0)^2 + \sum 1/2\ K(1+\cos 3\phi)$$

$$+ \sum [A\exp(-Br) - C/r^6]$$

Table 8.6 Parameters for glucose

type	K	b0,θ0
C-C	720	1.52
C-O	863	1.42
C-H	720	1.09
O-H	806	0.97
C-C-C	143.9	all =
C-C-O	143.9	1.911
C-C-H	93.5	
C-O-C	143.9	
C-O-H	80.6	
O-C-O	143.9	
O-C-H	93.5	
H-C-H	74.8	
X-C-C-X	2.40	
X-C-O-X	1.54	

	A*10 (-4)	B	C
C---C	23.70	4.32	297.8
C---O	21.21	4.44	244.0
C---H	3.14	4.20	121.1
O---O	18.64	4.55	200.0
O---H	2.81	4.32	99.2
H---H	0.66	4.08	49.2

X = C, O, H

9 REFERENCES

Adachi, N. (1971) J. Optim. Theor. Appl. 7: 391.

Allen, F.H. and Rogers, D. (1969) Acta Crystallogr. B 25: 1326.

Allinger, N.L., Tribble, M.T., Miller, M.A. and Wertz, D.H., (1971)
    J. Am. Chem. Soc. 93: 1637.

Allinger, N.L. (1976) Adv. Phys. Org. Chem. 13: 1.

Altona, C. and Faber, D. (1974) Fortschr. Chem. Forsch. 45: 1.

Altona, C. and Sundaralingam, M. (1970) J. Am. Chem. Soc.
    92: 1995.

Beveridge, G.S. and Schechter, R.S. (1970) Optimization: Theory
    and Practice, McGraw-Hill, New York.

Birshtein, T.M. and Ptitsyn, O.B. (1966) Conformations of Macro-
    molecules, Wiley-Interscience, New York.

Bixon, M. and Lifson, S. (1967) Tetrahedron 23: 769.

Blackburne, I.D., Duke, R.P., Jones, R.A.Y., Katritzky, A.R. and
    Record, K.A.F. (1973) J. Chem. Soc. Perkin II 332.

Box, M.J. (1966) Comput. J. 9: 67.

Boyd, R.H. (1968) J. Chem. Phys. 49: 2574.

Branin, F.H. (1972) IBM J. Res. Develop. 16: 504.

Branin, F.H. and Hoo, S.K. (1972) in Numerical Methods for Non-
    Linear Optimization, F.A. Lotsma (Ed.), Academic Press,
    London, p. 231.

Brent, R.P. (1973) Algorithms for Minimization without Derivatives,
    Prentice-Hall, Englewood Cliffs, New Jersey.

Buckingham, R.A. (1958) Trans. Faraday Soc. 54: 453.

Buckingham, D.A. and Sargeson, A.M. (1971) Topics in Stereochem.
    6: 219.

Cahn, R.S., Ingold, C. and Prelog, V. (1966) Angew. Chem.
    Internat. Edit. Engl. 5: 385.

Cauchy, A. (1847) Compt. rend. Sci. (Paris) 25: 536.

Clementi, E. and van Niessen, W. (1971) J. Chem. Phys. 54: 521.

Corey, E.J. (1971) Quart. Rev. (London) 25: 455.

Corey, E.J. and Wipke, W.T. (1969) Science 166: 178.

Coulson, C.A. and Danielsson, U. (1954) Arkiv Fysik 8: 239, 245.

Davidon, W.C. (1959) AEC Research and Development Report,
    ANL-5990 (Rev.).

De Coen, J.L., Elefante, G., Liquori, A.M. and Damiani, A. (1967)
    Nature 216: 910.

Del Re, G. (1958) J. Chem. Soc. : 4031.

Del Re, G., Pullman, B. and Yonezawa, T. (1963) Biochim. Biophys.
    Acta 75: 153.

Dugundji, J. and Ugi, I. (1973) Fortschr. Chem. Forsch. 39: 19.

Dunitz, J.D., Eser, H., Bixon, M. and Lifson, S. (1967)
    Helv. Chim. Acta 50: 1572.

Dwyer, M. and Searle, G.H. (1972) J.C.S. Chem. Commun. 726.

Eliel, E.L., Allinger, N.L., Angyal, S.J. and Morrison, G.A.
    (1965) Conformational Analysis, Wiley-Interscience, New York.

Engler, E.M., Andose, J.D. and Schleyer, P.v.R. (1973)
    J. Am. Chem. Soc. 95: 8005.

Ermer, O. and Lifson, S. (1973) J. Am. Chem. Soc. 95: 4121.

Ermer, O. and Lifson, S. (1974) J. Mol. Spectr. 51: 261.

Ermer, O. (1974) Tetrahedron 30: 3103.

Ermer, O. (1975) Tetrahedron 31: 1849.

Ermer, O. (1976) Calculation of Molecular Properties Using Force
    Fields. Application in Organic Chemistry. Structure and
    Bonding 27: 161.

Eyring, H. (1932) Phys. Rev. 39: 746.

Fiacco, A.V. and McCormick, G.P. (1968) Non-Linear Programming: Sequential Unconstrained Minimisation Techniques, Wiley, New York.

Fletcher, R. and Powell, M.J.D. (1963) Comput. J. 6: 163.

Fletcher, R. and Reeves, C.M. (1964) Comput. J. 7: 149.

Fletcher, R. (1965) Comput. J. 8: 33.

Flory, P.J. (1969) Statistical Mechanics of Chain Molecules, Wiley-Interscience, New York.

Fuehrer, H., Kartha, V.B., Krueger, P.J., Mantsch, H.H. and Jones, R.N. (1972) Chem. Rev. 72: 439.

Gans, P. (1976) Coord. Chem. Rev. 19: 99.

Gasteiger, J., Gillespie, P.D., Marquarding, D. and Ugi, I. (1974) Fortschr. Chem. Forsch. 48: 1.

Gelin, B.R. and Karplus, M. (1975) J. Am. Chem. Soc. 97: 6996.

Gentleman, W.M. (1973) J. Inst. Maths. Applics. 12: 329.

Gibson, K. and Scheraga, H.A. (1967) Proc. Natl. Acad. Sci. U.S. 58: 420.

Gill, P.E., Murray, W. and Picken, S.M. (1972) Natl. Phys. Lab. Report NAC 24.

Gleicher, G.J. and Schleyer, P.v.R. (1967) J. Am. Chem. Soc. 89: 582.

Go, N. and Scheraga, H.A. (1973) Macromolecules 6: 525.

Goldfeld, S.M., Quandt, R.E. and Trotter, H.F. (1966) Econometrica 34: 541.

Gollogly, J.R. and Hawkins, C.J. (1969) Inorg. Chem. 8: 1168.

Gordon, A.J. and Ford, R.A. (1972) The Chemist's Companion, Wiley-Interscience, New York.

Gourlay, A.R. and Watson, G.A. (1973) Computational Methods for Matrix Eigenproblems. Wiley, London.

Gregory, R.T. and Karney, D.L. (1969) A Collection of Matrices
    for Testing Computational Algorithms, Wiley, New York.

Gwinn, W.D. (1971) J. Chem. Phys. 55: 477.

Hagler, A.T. and Lifson, S. (1974) Acta Cryst. B30: 1336.

Hagler, A.T., Huler, E. and Lifson, S. (1974) J. Am. Chem. Soc.
    96: 5319.

Hagler, A.T. and Lifson, S. (to be published) Calculation of
    Protein Conformations, in The Proteins, H. Neurath (Ed.)
    3rd Ed., Vol. 5, Academic Press, New York.

Harary, F. (1969) Graph Theory, Addison-Wesley, Reading, Mass.

Harris, H.A. (1966) Ph. D. Thesis, Yale University.

Hendrickson, J.B. (1971) J. Am. Chem. Soc. 93: 6847, 6854.

Hirschfelder, J.O., Curtiss, C.F. and Bird, R.B. (1954) The Mole-
    cular Theory of Gasses and Liquids, Wiley, New York.

Hooke, R. and Jeeves, T.A. (1961) J. Assoc. Comput. Mach.
    8: 212.

Hopfinger, A.J. (1973) Conformational Properties of Macromolecules.
    Academic Press, New York.

Huang, H.Y. (1970) J. Optim. Theor. Appl. 5: 405.

Huang, H.Y. and Levy, A.V. (1970) J. Optim. Theor. Appl. 6: 269.

Hudson, B., Warshel, A. and Gordon, R.G. (1974) J. Chem. Phys.
    61: 2929.

Huler, E. and Warshel, A. (1974) Acta Cryst. B30: 1822.

Jacob, J., Thompson, H.B. and Bartell, L.S. (1967) J. Chem.
    Phys. 47: 3736.

Jacoby, S., Kowalik, J. and Pizzo, K. (1972) Iterative Methods
for Nonlinear Optimisation Problems, Prentice-Hall,
Englewood Cliffs, New Jersey.

Johnson, C.K. (1965) ORTEP: A FORTRAN thermal-ellipsoid plot
program for crystal structure illustrations, ORNL-3794
(Revised), Oak Ridge National Laberatory, Oak Ridge,
Tennessee.

Kildeby, K., Melberg, S. and Rasmussen, Kj. (1977) Acta Chem.
Scand. A31: 1.

Kim, P.H. (1960) J. Phys. Soc. Japan 15: 445.

Kowalik, J. and Osborne, M.R. (1968) Methods for Unconstrained
Optimisation Problems, Elsevier, New York - London - Amsterdam.

Lennard-Jones, J.E. (1931) Proc. Roy. Soc. 43: 461.

Levitt, M. (1971) Ph.D. Thesis, University of Cambridge.

Levitt, M. and Lifson, S. (1969) J. Mol. Biol. 46: 269.

Lewis, P.N., Momany, F.A. and Scheraga, H.A. (1973) Israel J.
Chem. 11: 121.

Lifson, S. and Roig, A. (1961) J. Chem. Phys. 34: 1963.

Lifson, S. and Zimm, B. (1963) Biopolymers 1: 15.

Lifson, S. (1963) Biopolymers 1: 25.

Lifson, S. (1964) J. Chem. Phys. 40: 3705.

Lifson, S. and Warshel, A. (1968) J. Chem. Phys. 49: 5116.

Lifson, S. (1968) J. Chim. Phys. Physicochim. Biol. 65: 40.

Lifson, S. (1972) Molecular Forces, in Protein-Protein Interactions, R. Jaenicke and E. Helmreich (Eds.), Springer-
Verlag, Berlin - Heidelberg - New York, p. 3.

Lifson, S. (1973) Recent Developments in the Consistent Force
    Field Calculations, in Dynamic Aspects of Conformation
    Changes in Biological Macromolecules, G. Sadron (Ed.),
    D. Reidel, Dordrecht, Holland, p. 421.

Liquori, A.M., Damiani, A. and Elefante, G. (1968) J. Mol. Biol.
    33: 439.

Liquori, A.M. (1969) Eleventh Nobel Symposium on Symmetry and
    Function of Biological Systems at the Macromolecular Level,
    A. Engström and B. Strandberg (Eds.), Almquist and Wiksell,
    Stockholm, p. 101.

London, F. (1937) Trans. Faraday Soc. 33: 8.

Lowe, J.P. (1969) Progress Phys. Org. Chem. 6: 1.

Lynch, M.F. (1968) Endeavour 27: 68.

Lynch, M.F., Harrison, J.M., Town, W.G. and Ash, J.E. (1972)
    Computer Handling of Chemical Structure Information,
    Macdonald, London.

Marquardt, D.W. (1963) SIAM J. 11: 431.

Marshall, C.W. (1971) Applied Graph Theory, Wiley, New York.

Matthews, A. and Davies, D. (1971) Comput. J. 14: 293.

McCormick, G.P. (1972) in Numerical Methods for Non-Linear
    Optimisation, F.A. Lotsma (Ed.), Academic Press,
    London, P. 209.

McGuire, R.F., Momany, F.A. and Scheraga, H.A. (1972) J. Phys.
    Chem. 76: 375.

Morse, P.M. (1929) Phys. Rev. 34: 57.

Murray, W. (1972) in Numerical Methods for Unconstrained
    Optimisation, W. Murray (Ed.), Academic Press, London.

Myers, G.E. (1968) J. Optim. Theor. Appl. 2: 209.

Nakagawa, I. and Shimanouchi, T. (1966) Spectrochim. Acta
    22: 759, 1707.

Nelder, J.A. and Mead, R. (1965) Comput. J. 7: 308; and Errata
      ibid. 8: 27.

Nemethy, G. and Scheraga, H.A. (1965) Biopolymers 3: 155.

Niketić, S.R. and Woldbye, F. (1973) Acta Chem. Scand.
      27: 621. 3811.

Niketić, S.R. and Woldbye, F. (1974) Acta Chem. Scand. A28: 248.

Niketić, S.R., Rasmussen, Kj., Woldbye, F. and Lifson, S. (1976)
      Acta Chem. Scand. A30: 485.

Nomenclature of Inorganic Chemistry (1971) Second Edition,
      Butterworths, London.

Orville-Thomas, W.J. (1974) (Ed.) Internal Rotation in Molecules.
      Wiley, London.

Pearson, J.D. (1969) Comput. J. 12: 171.

Pethrik, R.A. and Wyn-Jones, E. (1969) Quart. Rev. (London)
      23: 301.

Pitzer, K.S. (1959) Adv. Chem. Phys. 2: 59.

Poland, D.C. and Scheraga, H.A. (1967) Biochemistry 6: 3719.

Powell, M.J.D. (1964) Comput. J. 7: 155, 303.

Ramachandran, G.N., Ramakrishnan, C. and Sasisekharan, V. (1963)
      J. Mol. Biol. 7: 95.

Ramachandran, G.N. and Sasisekharan, V. (1968) Adv. Protein
      Chem. 23: 283.

Ramachandran, G.N. and Srinivasan, R. (1969) Int. J. Protein
      Res. 1: 5.

Ramachandran, G.N. and Srinivasan, R. (1970) Indian J. Biochem.
      7: 95.

Ramachandran, G.N., Venkatachalam, C.M. and Krimm, S. (1966)
      Biophys. J. 6: 849.

Ramakrishnan, C. and Ramachandran, G.N. (1965) Biophys. J. 5: 909.

Rosen, J. (1964) Brown Univ. Comput. Rev. 1: 64.

Rosenbrock, H.H. (1960) Comput. J. 3: 175.

Scheraga, H.A. (1971) Chem. Rev. 71: 195.

Schellman, J. and Lifson, S. (1973) Biopolymers 12: 315.

Schlessinger, J. and Warshel, A. (1974) Chem. Phys. Lett. 28: 380.

Scott, D.W., Messerly, J.F., Todd, S.S., Guthie, G.B., Hosenlopp,
    I.A., Moore, R.T., Osborn, A., Berg, W.T. and McCullough,
    J.P. (1961) J. Phys. Chem. 65: 1320.

Shah, B.V., Buehler, R.J. and Kemphtorne, O. (1964) SIAM J. 12: 74.

Simanouti, T. (1949) J. Chem. Phys. 17: 245, 734, 848.

Smith, C.S. (1962) Natl. Coal Board Sci. Dept. Report
    SC 846/MB/40, London.

Smith, E.G. (1968) The Wiswesser Line-Formula Chemical Notation,
    McGraw-Hill, New York.

Sovers, O.J., Kern, C.W., Pitzer, R.M. and Karplus, M. (1968)
    J. Chem. Phys. 49: 2592.

Stolow, R.D. (1971) in Conformational Analysis: Scope and Present
    Limitations, G. Chiurdoglu (Ed.), Academic Press, New York.

Sutton, L.E. (1965) Tables of Interatomic Distances, Supplement
    Special Publ. No. 18, The Chemical Society, London.

Tamburini, B., Tristo, G. and Del Pra, A. (1973) J. Chem. Phys.
    59: 3105.

Torrens, I.M. (1972) Interatomic Potentials, Academic Press,
    New York.

Warshel, A. and Lifson, S. (1969) Chem. Phys. Lett. 4: 255.

Warshel, A. and Lifson, S. (1970) J. Chem. Phys. 53: 582.

Warshel, A. (1971) J. Chem. Phys. 55: 3327.

Warshel, A. Levitt, M. and Lifson, S. (1970) J. Mol. Spectr.
    33: 84.

Warshel, A. and Karplus, M. (1972) J. Am. Chem. Soc. 94: 5612.

Warshel, A. and Karplus, M. (1974) J. Am. Chem. Soc. 96: 5677.

Warshel, A. (1973) Israel J. Chem. 11: 709.

Warshel, A. (1977) The Consistent Force Field and its Quantum
    Mechanical Extension, in Modern Theoretical Chemistry,
    Vol. 7, G.A. Segal (Ed.), Plenum Press, New York.

Whiffen, D.H. (1976) in Faraday Disc. No. 62, The Chemical
    Society (in print).

Wiberg, K.B. (1965) J. Am. Chem. Soc. 87: 1970.

Wilkinson, J.H. The Algebraic Eigenvalue Problem. Clarendon
    Press, Oxford 1965.

Williams, J.E., Stang, P.J. and Schleyer, P.v.R. (1968)
    Ann. Rev. Phys. Chem. 19: 531.

Wilson, E.B. (1959) Adv. Chem. Phys. 2: 367.

SUBJECT INDEX